ACHIEVING ENERGY INDEPENDENCE - ONE STEP AT A TIME

by

JEFFREY R. YAGO, P.E., C.E.M.

Printed in the United States of America
ISBN: 0-9669336-0-5
Library of Congress Catalog Card Number: 99-90121

Second Printing: February 2001

Quantity book orders are available for training or dealer programs by contacting the Publisher:

DUNIMIS TECHNOLOGY INC.
Post Office Box 10
Gum Spring, Virginia 23065
Phone: 804-784-0063

NOTICE TO READERS

The material in this document is for voluntary acceptance and use by the reader. This document does not define wiring or safety standards for system design or construction. Any included design guidelines, data tables, or "rule of thumb" information is intended to assist the reader in developing an overall project scope and preliminary cost estimate. All wiring, wiring devices, and electrical equipment installed in new or existing construction is subject to local building codes, National Construction Standards, and the National Electric Code. All wiring should be installed by licenced electricians. This material is subject to revision as further experience and product developments may deem necessary in this rapidly changing field.

Any products or equipment referred to by brand name or model number is for informational purposes only, and is not intended to be an endorsement by the author.

ACKNOWLEDGMENTS

I would like to thank Windy Dankoff with Dankoff Solar, Jan Klima with CoSEIA, Bob Fore with Heart Interface, Jim Kron with Heliotrope General, Jim Staber with Staber Industries, Helga Canfield with Southwest WindPower, Larry Schlussler with SunFrost, Bill Hollibaugh with Holly Solar Products, David Rippner with Applied Power, Jim Drizos with Trojan Batteries, Tobin Booth and Greg Thomas with Trace Engineering, and Donald Warfield with BP Solarex for their assistance in developing the original manuscript.

It would be nice to say I wrote this book, but in reality it wrote itself. It took the understanding of many clients who allowed those of us involved since the early solar beginnings of the 1970's to learn by our mistakes, since books like this did not exist. It took the solar equipment manufactures dedicated to building truly reliable products for homeowners in life threatening situations if there equipment failed. It took the many installers and distributors of solar photovoltaic systems who were willing to share their experiences and knowledge for the benefit of us all. And finally, this book could have never been written without the dedication, computer skills, and untiring efforts of my wife and co-author, Sharon Seymore Yago.

Thanks to all! Jeffrey R. Yago

TABLE OF CONTENTS

ACHIEVING ENERGY INDEPENDENCE - ONE STEP AT A TIME

INTRODUCTION

Figure 1.1 The author's Solar Powered Residence in Virginia. System includes passive solar greenhouse, hydronic wood stove, 2 kW photovoltaic solar array, and backup propane fueled 6 kW generator. Home was awarded 1st Place Low Energy Design 1992 Virginia Department of Energy, 1st Place Low Energy Design 1994 Virginia Propane Dealer's Association, and was selected one of ten finalists in National Gallery of Architectural Design Competition. This 3,400 square foot home averages $33 per month in utility costs.

My days usually begin with a call like this: "I'm getting concerned about living in the city and I am buying land in an isolated area for a retirement home. I want all the modern conveniences that I'm accustomed to, however; the utility power is subject to long outages or my property is not near a utility line." "I am concerned that the electric grid may fail and who know how long it will take to resolve the problem." "I live in California and rolling blackouts have shut down my business phones, computers, and web site."

Are you considering a life style change, moving to a rural area, building a vacation home where utility power is not available, or have decided you no longer want to be totally dependent on others to provide all of your energy needs? Do you have a home based business and need backup power for your computer and fax machine? Perhaps you already live in an area where storms cause week long power outages and this inconvenience is becoming all too commonplace. Are headlines about major electric and gas utilities going bankrupt or experiencing shortages making you feel vulnerable because you are totally dependent on something you cannot control?

Experts agree that the electric grid bringing power into New York and New England from nearby power exporting states has already reached its original design capacity and these areas may soon experience California style blackouts, but for different reasons.

Let me start by saying you are not alone and your numbers are growing. This text will serve as your guide through the sometimes confusing world of alternative electrical power systems. By taking this "one step at a time," you can gradually achieve energy independence and the peace of mind that it brings. We will begin with a brief discussion on the utility grid, then review where electricity is used in your own home or office. A very strong emphasis will be placed on high efficiency lighting and appliances first, as this will greatly reduce the size and cost of any backup power system you may decide to install.

This material has been divided into chapters having the most basic information presented first. I realize some readers may have a strong technical background and want to skip over these first few chapters; however, I think everyone will find something of interest in this review. Later chapters will build on these basics by describing electrical loads in more detail, and present ways to reduce these loads. The final chapters discuss how your electrical loads can be powered from generators or alternative energy systems, identify which systems are right for your unique situation, and provide helpful preliminary wire and component sizing procedures.

CHAPTER I
WHAT IS THE GRID?

Most people today have little understanding of electricity, or how it is distributed. Unlike an alternative energy system which has a given total capacity and specific period of time power can be provided, today's electrical grid can deliver to your wall outlets all the power you could ever use, and for as long as you want it (assuming you can pay the bill). However, for this unlimited energy to reach our wall outlets, there is an orchestrated power flow ballet in progress at all times between the power generating plants and the millions of electrical users which must be carefully balanced and managed or the whole house of cards would collapse.

When Thomas Edison built the first power station at his headquarters in Menlo Park, New Jersey in 1879, it consisted of three steam driven generators which powered a string of 100 carbon filament street lamps and laboratory building lighting. The plant operators knew when the lights would be turned on and off, and the plant could be operated based on this specific schedule and constant electrical demand. The Pearl Street Station in New York City was the first commercial central electric generation plant. It was completed by Thomas Edison in 1882 and was designed to provide lighting to one square mile of the city and used underground electrical conduits. Soon larger areas became electrified, additional generators were added at different locations and wired to the same distribution system to help reduce line losses, thus creating the first grid of interconnected generating plants and system loads.

Just prior to this installation, the New York Board of Fire Underwriters adopted in October 1881, seven rules to follow before applying for permission to install these electric lights in any city building. During the next five years the National Board of Fire Underwriters met with other state agencies and equipment manufacturers to review this introduction of electrical technology into buildings. This resulted in the first National Electrical Code being published in 1897, over 100 years ago. Today, all residences and commercial electrical wiring, including solar photovoltaic systems, must be installed according to the National Electric Code.

The first metered electric bill was dated January 1883 for $50.40 for power to a New York manufacturing plant and by 1883 this first commercial electric grid was supplying over 10,000 electric lights serving 500 customers. The first electric motor loads were added in 1884. The predictable loads of the first street lighting system had become a large number of homes and businesses, with lights and motor driven machinery turning on and off at all hours of the day. During factory work days all generators needed to operate at full capacity, yet a few hours later many could be shut down.

Additional metering systems and power operators needed to be added to bring some order to the chaos and to help predict when the uncontrolled loads were about to change so generating capacity was ready and available. If the operators guessed wrong, a lack of generating capacity during rapidly increasing loads would cause the grid voltage to drop, just as operating more faucets or showers at the same time in your home will cause the water pressure to drop and all faucets receive less water flow. This is usually referred to as a "brown out". When the reverse occurs and a rapid reduction in electric usage occurs while all generators are operating at full capacity, a voltage increase can occur which can damage motors and burn out light bulbs.

If you have ever operated a water faucet at full flow then quickly shut the valve, you might hear a loud thump called "water hammer," as the large mass of moving water instantly comes to a stop with no place to go. The same problem can occur in an electrical grid as operators are required to switch from generator to generator and sub- station to sub-station to maintain system balance. When an electric circuit under heavy load is rapidly switched off, arcs occur across the switch contacts as the suddenly blocked current flow tries to keep going. This rapid switching of power flow can introduce a momentary voltage peak or "spike", which can cause extreme damage to delicate electronic components and appliances even through this spike occurs for only a very small fraction of a single second. Lighting strikes in distant locations can travel down the electrical grid and into our homes which can also create a rapid voltage rise and destroy stereos and televisions.

In actuality, the electric grid and the conveniences it provides really is a bargain - when its on. Less than thirty years ago a power outage was only a minor inconvenience to most households. As late as the mid-fifties, homes were not air conditioned and televisions were small, black and white novelties. Stoves, space heating equipment, and hot water heaters were gas or oil fired and their simple temperature controls did not require electricity or micro-electronics to operate. When the lights went out everyone brought out the flashlights and candles or went to bed. At that time only a small percentage of our population lived in highrise apartments, which, unlike houses, need more access lighting and motor driven elevators. Today, many highrise apartments and office buildings have non-opening windows which demands constant heating or cooling by electrical driven equipment to maintain comfortable room temperatures.

We are becoming more and more dependent on the electric grid, at the same time our electric appliances are becoming more sensitive to electrical power problems. A toaster or incandescent light operating in the 1960's could easily withstand power interruptions, brown outs, voltage spikes, utility drop-outs, and harmonics with little more than a brief pause in operation until the power came back on. Today, almost everything electrical requires micro-electronic components and controls which can be literally "fried" by these common power related problems.

As more and more computers, lighting ballasts, variable speed motor drives, and telecommunications equipment are added to a building load, more electric "noise" is imposed back into the electrical system which can also affect other electronic equipment. To further complicate this already complex system of generating plants and power distributions substations, the pending deregulation of the entire electric industry will bring even more pressure to bear on this delicate balance between power generation and system loads.

Large factories will be switching from one electric supplier to another depending on time of day, weather, or season. Some electric suppliers will have low nighttime electric costs and high mid-day charges, while others may have a totally different rate structure. Look at the deregulation of the telephone industry, we will soon have totally separate and competing corporations independently responsible for each power plant, the high voltage distribution lines, the local utility grid, and customer metering.

Will this mean more power outages as separate companies argue over who is responsible for replacing a burned out transformer? Will smaller and potentially less reliable independent power producers be off line when needed during peak periods causing more brown outs? As the existing electrical system ages, will competition cause operators to defer preventative maintenance to keep stockholders profits high at the expense of end users experiencing more power interruptions? Will the almost daily increases in government regulations, environmental restrictions, and fuel price fluctuations force older or less efficient power producers out of the market, leaving us all more dependent on fewer and fewer sources of electrical energy generation?

Concerns that December 31, 1999 would bring massive Y2K related system failures did not materialize. This non-event was due in large part to extensive software and hardware upgrades that were completed months prior to the date change. This should not however, make us feel there never was a problem, or that the utility grid is not vulnerable to failure resulting from computer hacker attacks or terrorist activities.

Once a system as complex and interconnected as the national electric grid goes down, you cannot just turn it back on with the flick of a switch. Unlike the manually operated grid of Edison's era, the control and regulation of the voltage and current on a nationwide interconnected grid of coal fired, gas fired, hydroelectric, and nuclear power plants requires a vast network of interconnected computer systems. Thousands and thousands of remote metering devices report under and over voltage conditions, high current demands, and excess system capacity. This allows computerized forecasting of when any power plant needs to be added or taken off line. Since it can take from four hours to an entire day to bring a large coal fired plant on line, this computerized load projection is critical.

When a plant is unexpectedly taken off line for repairs, other plants are adjusted to pick up the slack. But what if there is a "trigger event" that causes multiple plants to shut down at the same time? Once this process begins, if not caught in time, other perfectly good generator plants must be shut down to avoid extreme equipment damage due to grid wide under voltage conditions causing high current flows.

The answer to all this is simply - no one really knows, but everyone agrees we could be in for a period of increased risk until these issues are resolved. Our total dependence on a very complex power distribution network, the increasing use of delicate micro-electronics which must have a highly stable electrical power source, and the uncertainties that utility deregulation and environmental restrictions will bring, should tell us our feelings of uncertainty are well founded.

Most power outages today are storm related and are easily corrected by repairing a downed power line. When major storms cause massive power outages, the most critical lines and customers are repaired first. This would include hospitals, airports, central city areas, and schools. Moving outward from the damaged primary feeders, power lines serving less critical areas are repaired next followed by smaller blocks of residential customers. Power lines serving more rural areas are usually last to be repaired which could take weeks in the aftermath of a major hurricane or tornado.

As more and more people start home based businesses or tele-commute from rural residential areas, they will be working in areas where power outages are more common and last longer due to repair delays. The potential financial impact on a home based business from not being able to make or receive phone calls, faxes, e-mails, or use a computer for days could be significant.

Now that you have a better understanding of the source of your home's electrical power, it is important to identify your own electrical power needs and how to satisfy these needs more efficiently. After you have determined what your electric needs are, we will discuss ways to minimize these loads before learning how to select the best backup or alternative energy system to power these loads.

CHAPTER II
IDENTIFYING YOUR POWER NEEDS

Just as we have a first aid kit for accidents, and fire extinguishers and smoke alarms for life safety, it's time we considered a backup power system as normal and prudent power insurance in today's world. Which of the lights around us are absolutely necessary in an emergency or power outage and what appliances, office equipment, or entertainment systems do we need to keep operating. Once you have identified your electrical necessities, the next step is to determine how long these systems and appliances must continue to operate without the utility grid. Some of you are thinking in terms of hours, while others may be thinking days, weeks, or forever if the utility grid does not serve your location.

At this point it is still too early to be thinking about the components of your planned alternative energy system. Most likely you are already discussing what generator to buy, do you need batteries, will solar modules work in your area, are wood stoves safe, do I need a wind generator or water wheel? This is like shopping for transportation and worrying about what tires to buy before deciding if a bicycle, car, truck, or bus will meet your transportation needs. You must first decide what you want an alternative energy system to power and for how long. This will be our starting point, and this will tell us what path to take. Ultimately we will discuss all major energy system components, but we will begin with the electrical powered lighting and small appliances as these are the easiest loads to identify and understand.

Put down this book for a moment and look around. What lights are on right now where you are reading and what appliances are operating at this moment? Are lights on in unoccupied rooms? Is a TV or stereo playing? Is an air conditioner operating or is a furnace blowing warm air? Are you having a cold drink with ice cooled by a freezer, or hot coffee heated by an electric brewer? Did you just microwave some popcorn while your desktop computer remains on to receive an e-mail and your answering machine just told another salesman you were not in?

The electrical energy an appliance consumes is measured in "watts" and this value is usually indicated on the appliance nameplate. Due to the small size of a watt, total electrical loads are given in units of kilowatts (1,000 watts). The consumption of 1,000 watts during any one hour period is called a kilowatt-hour (kWh).

In order to properly size a backup or alternative energy system we need to solve two problems: What is the total energy the system must deliver over a fixed time period (kilowatt-hours), and what is the peak load the system must meet (kilowatt). I find these two problems are easier to manage if we divide the 24-hour day into blocks of time. Since any time period can be calculated, we are looking for the time of day when a higher number of lights and appliances would be operating at the same time.

For a typical household or office, one of the highest demand periods occur between 5 AM to 8 AM when people are waking up and getting ready for work. Although this peak load may last less than an hour, we might expect to find a well pump supplying several showers, a hot water tank heating water, someone plugging in a coffee maker and toaster while another person is operating a hair dryer or curling iron, the radio or television is delivering the morning news, while the majority of bedroom, bathroom, and kitchen lights are on. It is doubtful, however, that we would also find the dishwasher, clothes dryer, air conditioner, attic exhaust fan, or outdoor yard lights operating at this early time of day. By 8:00 or 9:00 AM most offices and factories come alive with almost all lights, computers, and shop machinery starting up at almost the same time. We may also want to look at these simultaneous loads that would occur around noon for homes with young children not in school, or facilities having kitchens and cafeterias.

The next high load period will probably start around 6:00 PM to 7:00 PM when people return home, adjust the thermostat, go into the kitchen and use mixers, stoves, microwave ovens, disposals and dishwashers, while listening to the 6:00 PM news on a television or radio. We would probably find all lights on in the kitchen, dining room, living room, den, and front porch or garage. After 8:00 PM things usually start to settle down as far as electrical loads are concerned. This may be the time a home computer is surfing the Internet, someone may be operating the clothes washer and dryer, a clothes iron may be heating up to iron a dress or shirt for tomorrow's workday, and more than one television can be heard in the background. The air conditioner may be operating at this time, but we probably would not find the hair dryer, coffee pot or toaster operating this late.

APPLIANCE INVENTORY: Let's begin by completing TABLE #1 in the Appendix, which is an inventory of all electrical loads we consider necessary, and the hours per day each operate. Non-continuous loads such as a clothes washer, and shop or hobby tools should not be included in this preliminary analysis unless these are everyday necessities for your particular situation. These occasional loads will be considered later for those of you planning to be totally independent of the utility grid. There is a list of the average electrical consumption for all major appliances and office equipment in TABLE #3 of the Appendix which will be helpful in completing TABLE #1.

It is important to divide the operating hours for each appliance into time periods, since for most cases, not all loads will be operating at the same time. Some loads may only be on in the morning, while others are only on at night. I have divided a typical day into four time periods which should separate the more energy intensive time periods from the more passive periods. This should help identify where you need to reconsider the times some appliances are operated. The random use nature of most electrical loads, and the ability to shift operating times from a heavily loaded time of day to a less active time period allows a major reduction in backup system design capacity, and associated installation costs.

Referring to TABLE #1, if we have two 75 watt lights that operate in the bedroom from 7 AM to 8 AM, and 6 PM to 12 midnight, we would indicate a total load of 150 watts, with one hour in the 5 AM and 9AM column, four hours in the 5 PM to 10 PM column, and two hours in the 10 PM to 5 AM column, for a total of seven hours per day. A constant load such as a fax or answering machine would have four hours in the 5 AM to 9 AM column, eight hours in the 9 AM to 5 PM column, five hours in the 5 PM to 10 PM column, and seven hours in the 10 PM to 5 AM column for a total of 24 hours per day.

TABLE #1 also allows us to determine what the maximum kilowatt (kW) load demand will be on our future backup power system or generator, and the time period this peak load occurs. This data will be used to calculate the peak capacity of our backup system components. Try shifting the larger loads to other time periods to see how this can reduce the size of a generator. After TABLE #1 has been completed for all critical loads, carry this data forward and complete TABLE #2 by multiplying the lighting and appliance load in watts, multiplied by the operating hours for each time period which equals kilowatt-hours (kWh). This will be needed to calculate the storage capacity of a battery bank or the operating hours of a generator.

After completing TABLE #1 and TABLE #2 for all <u>critical</u> electrical loads, reconsider these loads and think energy conservation. Do you see lights or appliances that should be replaced? This will pay off significantly when it comes time to purchase a backup generator or off -grid power system.

Now that you have a preliminary idea as to the kW peak capacity that the backup or off-grid power system must meet, and the kWh energy consumption the system loads will have, it's time to review several possible system designs that will meet your requirements.

EFFECTS OF TIME ON ENERGY USAGE: Energy consumption is time based and equipment efficiencies are affected by the time period they supply a given peak load. For example, we can build a car to travel a total of 60 miles in a day on a half-gallon of gasoline. It is obvious this car must be very light weight and have a very small motor to produce these energy conserving results, but we will cover the 60 miles distance - eventually. If we want to travel the same distance in one hour, we will need to increase the motor and drive train capacity which will also increase the vehicle weight. We will travel the same 60 miles distance much faster, but our vehicle may now consume 2 gallons of gas.

Electrical energy works the same way. A microwave oven can heat food in minutes that would take hours in a crock pot. But the microwave oven has over 1,000 watts of electrical load as compared to a 200 watt crock pot or hot plate. Using the following example, let's say it takes 15 minutes to heat a pot of soup in a microwave oven. This consumes 250 watt-hours of energy or 0.25 kW.

$$\textbf{(1,000 watts)}\left(\frac{\textbf{15}}{\textbf{60}}\right) = \textbf{250 watt-hours} = \textbf{0.25 kWh}$$

Notice that watt hours requires the time given in minutes to be converted to hours by dividing by 60. Let's say our 200 watt crock pot takes 4 hours to heat the soup to the same temperature.

$$\textbf{(200 watts)(4)} = \textbf{800 watt-hours} = \textbf{0.8 kWh}$$

The crock pot or hot plate consumes much less energy per unit of time than the microwave; however, the microwave would be operating a shorter period of time which results in a lower total energy consumption.

If we were designing an emergency backup power system to also power an appliance to cook food, which system would work better? It is obvious the appliance that consumes the least total energy watt-hours is the most efficient and would have the smallest drain on a battery or generator. But what if your battery bank, wiring, or DC to AC inverter can only take a 500 watt load? In this case the microwave would exceed the current carrying capacity of the backup electrical system even though it would operate a fraction of the time required by the crock pot or hot plate.

The goal is to reduce the total electrical consumption and peak load by the combination of more efficient appliances and less loads operating at the same time.

AIR CONDITIONING LOADS: If you live in an area that requires air conditioning throughout most of the summer months, it would be very expensive to add enough solar modules to power a typical refrigerant compressor based central air conditioner or heat pump. You also may not want to operate a generator continuously to power a central air conditioning unit. A solar array large enough to power a central air conditioning system is usually over sized for the remainder of the year which reduces the cost effectiveness of a solar powered air conditioning system.

Many areas of the country could get by without cooling if the humidity level inside the home could be reduced. There are several new products on the market that use a desiccant material to absorb the moisture in re-circulated room air. Once the desiccant material is saturated, passive solar heat can be used to “dry out” the chemical for reuse. These chemicals usually have an infinite life and the fan or blower requires much less electricity to operate than a refrigeration compressor. This early air conditioning technology that was later replaced by the refrigerant compressor system may again become popular for air conditioning applications that must use minimum electricity.

If your present home includes a central air conditioning system, you will probably want to leave this system connected to the electric grid.

If you are planning a true off-grid home, it is much easier and less costly to heat with wood or propane than it is to cool with electricity. Therefore, it would be to your advantage if your planned off-grid home was located at a higher elevation or geographic area that has relatively mild summer temperatures or low humidity. The most cost effective air conditioning system available for a solar home is opening the window and operating a ceiling fan.

Figure 2 - 1 A high quality ceiling fan provides "free cooling" while operating quietly on minimum electricity.

Some western states have very hot and dry summers, which is ideal weather to operate an evaporative cooler. These rooftop mounted units consist of a filter pad and small water spray nozzle. Some units include an internal fan, while others utilize an exhaust fan located at the opposite end of the home. Dry outside air is drawn down through the moist filter pads where the high ambient temperatures easily evaporate the water vapor out of the air stream. The resulting air flow is cooler in temperature, but slightly higher in humidity than the original air stream; however, in many locations these "swamp coolers" are very effective and require only fan energy to operate.

LIGHTING LOADS: Many people are not aware that there has been tremendous improvements in the lighting fixture efficiency during the 1990's, and that there are many new types of energy efficient bulbs and tubes now available. In addition, several lamps are being legislated out of production including the 150 watt PAR 38 usually used as an outside flood lamp, and the 40 watt T-12 fluorescent tube found in older office and shop ceiling fixtures.

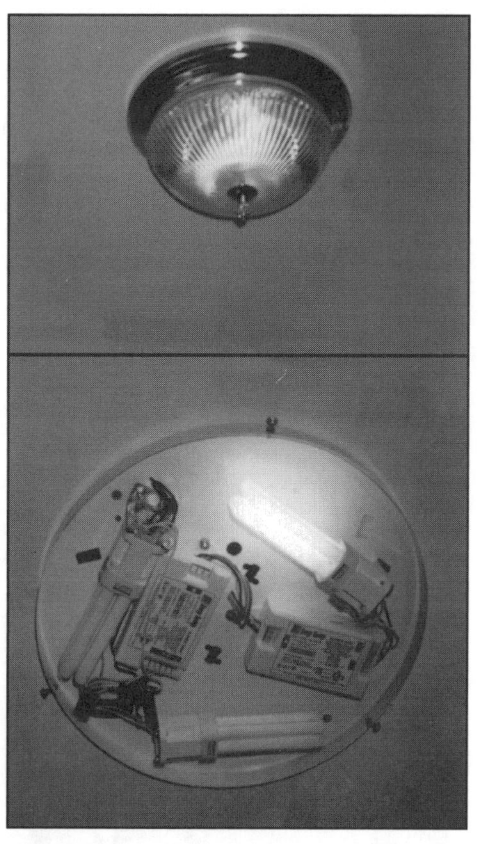

Figure 2-2 Decorative ceiling fixture using three energy saving compact fluorescent lamps with electronic ballists.

The most common replacement for a standard incandescent light bulb is the compact fluorescent lamp. Originally only available in the PL or "twin tube" configuration, many are now manufactured in the "quad" configuration with two separate pairs of tubes which allows the same wattage output with half the size. Unlike incandescent bulbs, all small fluorescent lamps require a ballast built into the tube base or mounted in the light fixture.

Figure 2-2 shows a very decorative ceiling fixture usually associated with three incandescent bulbs, that is now available with up to three compact fluorescent tubes and electronic ballasts. Although most quality PL or Quad style compact fluorescent lights work well with most inverters, there have been problems with several brands of fluorescent lights producing a noticeable audible hum when powered from inverter systems. It appears the electronics in the lamp ballasts are affected by the harmonics or non-sine wave output of some lower cost inverters.

The new 4-pin BIAX style compact fluorescent tube almost totally eliminates this problem, and the 4-pin design provides a warm-up delay which extends lamp life and improves operation in very cold ambient temperatures.

Most bedrooms up to 12 ft. x 12 ft. will require one ceiling fixture having a minimum of two compact fluorescent lamps. Each lamp must be at least 13 watts each to provide acceptable lighting levels. Larger rooms will require a minimum of three lamps or switch to higher wattage lamps. Rooms larger than 16 ft. x 16 ft. may require two ceiling fixtures equally spaced. Be sure all fluorescent lamps and tubes are ordered with a full spectrum or warm color temperature index.

Figure 2-3 shows the most common compact fluorescent fixtures. The self-ballasted Quad fixture on the left has a screw base and easily replaces an incandescent light bulb.

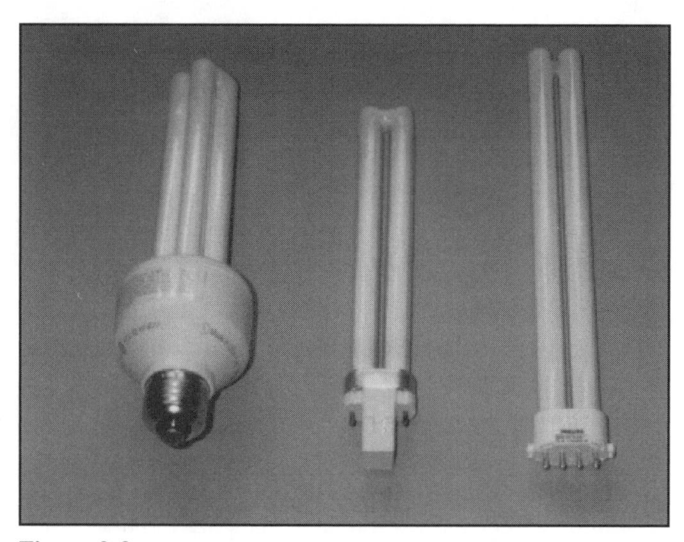

Figure 2-3 Most common types of compact fluorescent fixtures.

The 2-pin PL twin tube in the center is now used in many office and school fixtures including wall sconces, exit signs, and recessed ceiling down lights. Due to its popularity in commercial applications, the 2-pin PL lamp is readily available at fairly low cost. The BIAX 4-pin twin tube on the right, is newer lighting technology and offers flicker free, full spectrum lighting with minimum hum and maximum lamp life. All of these lamps make excellent replacements for incandescent bulbs, and are considered essential for an off-grid or low energy home.

Although some manufacturers now offer electronic ballasts that allow dimming fluorescent lamps, these have higher costs and can reduce lamp life. Many dimmer wall switches do not operate properly when connected to modified sine wave inverter based power systems.

To simplify your life, I suggest avoiding dimming lighting circuits unless you plan to use a very sophisticated inverter and high quality dimmer switches and ballasts.

TABLE 2-1 below provides the physical dimensions and light output for the most popular sizes of these lamps.

TABLE 2-1

Lamp Style (Watts)	Length	Lumens	Incandescent Replaced (Watts)
PL 7	5 ½”	400	40
PL 9	7 ½”	600	50
PL 13	9”	900	60
QUAD 13	4 ¾”	900	60
QUAD 18	6 ¾”	1,250	75
QUAD 26	7 ½”	1,800	120
BIAX 11	8 ½”	840	60
BIAX 18	10 ½”	1,250	75
BIAX 27	12 ¾”	1,800	120
BIAX 39	16 ½”	2,900	200

Many compact fluorescent lamps are now available with electronic ballasts that can operate from a 12 volt or 24 volt DC power source. Figure 2- 4 shows a wall sconce model using an 18 watt 4-pin BIAX compact fluorescent lamp. This makes an excellent wall sconce for corridors and stairs which can keep these critical access areas lighted during power outages if they are powered directly from a solar charged battery system.

For high ceiling applications and lighting fixtures controlled by a dimmer switch, a halogen bulb makes an excellent replacement for an incandescent flood lamp.

The halogen lamp will be more expensive, but it provides much longer lamp life and about $^{1}/_{3}$ the energy usage. Halogen lamps are also excellent over bathroom sinks or dressing tables, as they have excellent color rendition when using a mirror.

Changing to energy efficient lighting is one of the easiest methods to reduce your present utility billing, and is absolutely a necessity for any off-grid application.

Lights that combine tiny multiple white L.E.D. lamps is the most energy efficient lighting technology available. Although fixtures using this technology are still not bright enough to light up a room, they make excellent reading or task lamps for small cabins. These lights are extremely expensive, but since they do not require a filament or ballast, they have a 20 + years life expectancy and can operate for weeks from a small battery. Figure 2 - 5 shows a 12 volt DC L.E.D. lamp designed to retrofit into a desk lamp rewired for battery power.

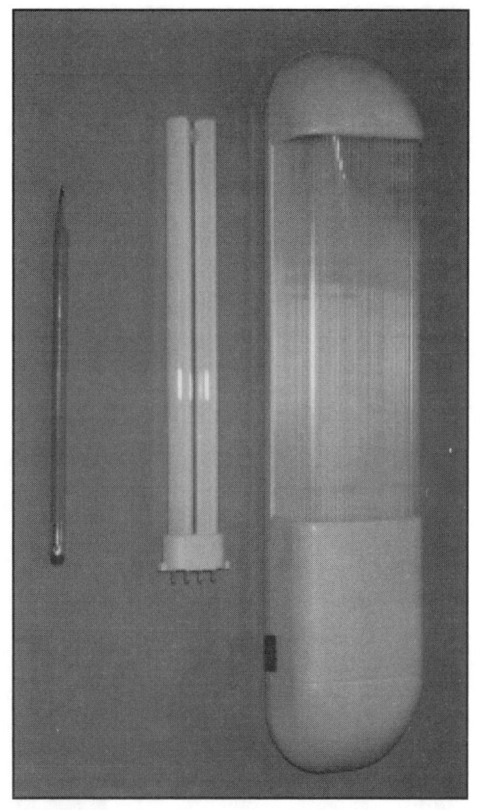

Figure 2-4 Very efficient 12/24 volt compact fluorescent wall sconce.

Figure 2-5 12 volt DC L.E.D. lamp.

HEATING APPLIANCE LOADS: Do not even think of using an electric heater or electric hot water tank if it will be powered by a generator or solar photovoltaic array. These appliances can consume all the energy from a fully charged residential battery bank in less than an hour. For new home construction in locations offering long hours of sun during the winter, passive solar space heating and solar hot water heaters are very effective and consume only minimum electricity to operate a small pump or circulating fan.

For existing homes or locations not suitable for solar heating, a propane fired hot water heater, kitchen stove, and backup hot water heating boiler are excellent choices.

If your space heating loads are not large, several domestic hot water heaters are available that can also provide hot water to baseboard radiation around the perimeter of the house, in addition to heating the domestic hot water.

Figure 2 - 6 shows an AquaStar wall mounted instantaneous hot water heater. These units are very popular for off-grid homes since they have a small size and weight, and do not require any electricity to operate. As soon as the hot water faucet is opened, a flow switch inside the heater switches the propane gas burner to full heat. A large copper and stainless steel coil around the

Figure 2 - 6 AquaStar tankless domestic hot water heater.

burner is designed to heat the water from ground temperatures up to the setpoint temperature without a storage tank.

Forced air furnaces are not recommended for off-grid homes, as the fan in any central air handling unit requires much more electrical energy to move heat throughout the house than required by a small circulating pump in a hot water heating system. Most central fans are $^{1}/_{2}$ HP, and may need to operate almost continuously on very cold days. This equals 5,000 watt hours of electrical energy per day based on twelve hours of fan run time per 24 hours.

If your system is intended for backup power only and you will use a conventional forced air furnace when utility power is available, then I suggest using a wood stove for heating during power outages. This will greatly reduce the electrical demand on your generator or battery bank.

CLOTHES WASHING AND DRYING: One of the most promising signs that U.S. appliance manufacturers finally "get the picture," is the Staber Industries Model 2300 clothes washer shown in Figure 2 -7 and 2 - 8.

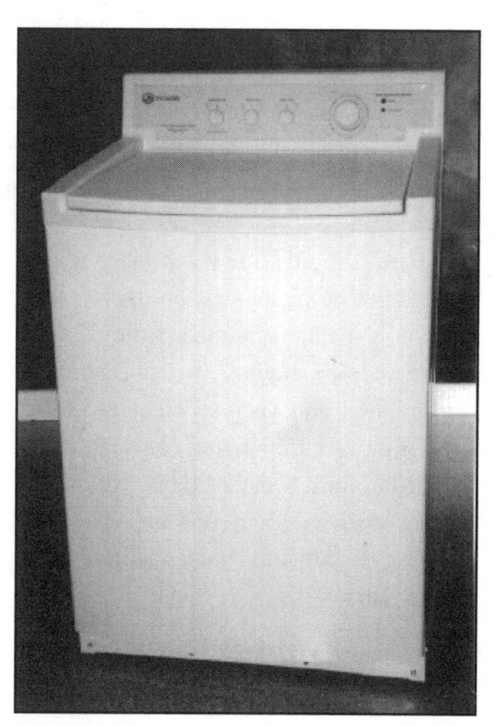

Figure 2 - 7 Energy efficient Staber 2300 clothes washer.

At almost double the cost of a commercial clothes washer, it may not fit the residential appliance market where cost is the first and only consideration. However, for an off-grid solar home, this washer offers many advantages including a unique top loading rotating drum design that does not require an agitator and associated gear boxes, drive shafts, and large drive motor.

This unit draws less than 800 watts with almost no starting in rush current, and will operate on an inverter as small as 1,000 watts.

It also uses only 21 gallons of water per wash cycle which is less than half the usage of a conventional washer. This also cuts the energy required to heat and pump this water by half which is a real savings for sites with water availability problems.

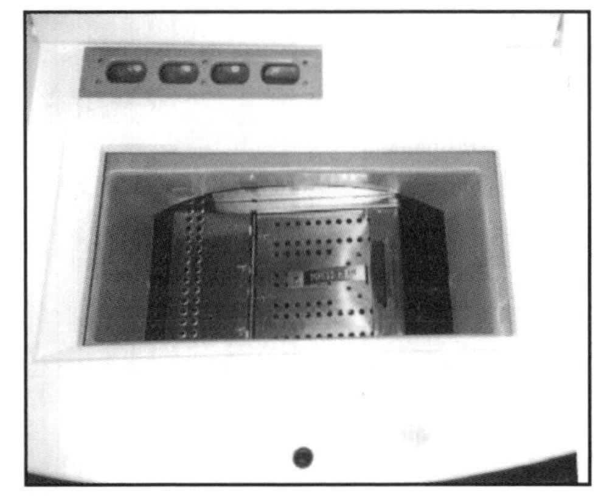

Figure 2 - 8 Close-up view showing top loading rotating drum.

A dryer may not be required for many homeowners, as the extremely high speed drying cycle removes most of the water remaining after the wash cycle. The rotating drum, fan belt, motor, and pump are all accessible from the removable front, and this attention to homeowner maintenance should be a real selling point for a remote site not accessible to the "Maytag" repairman.

There are still only a few appliance manufacturers in the United States that offer super efficient appliances, but there are several quality super efficient appliance lines now being imported from Europe. The SPLENDIDE line offers an apartment size clothes washer and vented dryer which has very low water consumption and very efficient electric drying. Although the dryer utilizes a 1,400 watt electric heating element, this is used in combination with an outside vented fan for a separate blow dry cycle to reduce energy usage to a minimum. Clothes washers and dryers by Asko and Creda are also very popular European imports for off-grid homes although they are smaller than their American counterparts.

CHAPTER III
PARASITIC ELECTRIC LOADS

We assume that when we turn off an appliance or stereo system the equipment is off; but nothing could be further from the truth. The satellite DSS receivers I have tested consumed the exact same power regardless of being turned on or off. Many small appliances and desk lamps utilize a transformer, called a "wall cube" that plugs into the wall outlet; however, this transformer is located between the power source and the appliance's on/off switch.

My testing using a calibrated digital wattmeter found these wall cube transformers to have a constant loss of 1 to 3 watts when the connected load was turned off. I also found that all component stereo equipment having remote controls averaged 3 to 7 watts loss while the equipment is turned off due to electronics that must remain operating to "hear" the remote on/off command. Kitchen appliances that include a digital clock timer were found to have a constant 3 watt loss when not in use.

A recently completed solar home I designed for an Idaho couple was requiring the backup generator to run 4 hours per day instead of the calculated 4 hours every three days.

After checking all systems we found that the owner installed a gas range that was using more electricity than the entire 2,500 square foot house including hydronic heating system! Further investigation found this "super efficient" GE stove required a 500 watt electric heating element to remain on as long as the gas oven was in use. Figure 3 - 1 provides a close-up view of this glowing element in operation.

Kitchen stoves and water heaters using electronic spark ignition instead of the ceramic heating element still were found to have a constant 5 to 10 watts per hour loss, and the 24 volt AC control transformers for doorbells, electric garage doors, and furnace thermostat controls each averaged 2 watts per hour loss.

The energy efficient design for an off-grid home should address these constant 24 hour per day parasitic electric loads, and gas fired appliances should be checked before purchase to insure they will operate with little or no electricity. To see how this can impact the cost of a solar powered home, the following example calculates these losses for a typical home and the appliances most likely in use.

Figure 3 - 1 Close-up view of gas oven burner and electric element ignition for GE gas stove.

EXAMPLE OF PARASITIC LOADS:

Quantity	Appliances	Total Watts
2	Halogen lamps with wall cube	2
1	Fax machine or cell phone charger	5
1	Cordless drill or screwdriver recharger	2
1	Answering machine	2
1	Cordless phone	2
1	Microwave oven with electronic clock	3
1	Stereo VCR tape player with remote	7
1	Electronic Ignition for gas appliance	5
1	Door bell transformer	2
1	CD disk player with remote	3
1	Stereo amplifier with remote	3
1	19” color TV with remote	4
		40 watts
		x 24 hours
		960 watt hours per day

Using a 50 watt solar photovoltaic module to collect energy for six hours of sun per day, would require over three solar modules supplying all of the energy they collect to make up these parasitic losses. This represents over $1,000 investment in photovoltaics and batteries just to supply this constant system drain. Keep in mind that none of these electrical losses are necessary, and most can be eliminated by better appliances selection. Note that although each individual electrical load is small, the constant 24-hour per day operation quickly adds up to a large drain on the battery power system. TABLE #3 in the Appendix includes the actual standby parasitic loads for all common home and home office equipment.

When selecting electrical appliances for an off-grid home, avoid equipment with remote controls and electronic clock type features. If this cannot be avoided, locate a double duplex wall outlet behind the proposed location of your stereo components and wire this outlet directly to a standard wall switch next to the room light switch. This will allow you to disconnect <u>all</u> power to this equipment when not in use. You can use the same strategy for your computer desk.

When selecting gas appliances for an off-grid home, remember that the old reliable pilot light is still available for most ovens and stoves but may require special order. Any small increase in gas usage by a pilot light is still less costly than the solar electricity that can be saved by not continuously powering electronic ignition controls.

CHAPTER IV
REFRIGERATION

Outside of air conditioning, electric refrigeration has the largest energy usage in a typical grid connected single family residence. The concept of mechanical cooling was first patented in 1834, but it was not until 1902 when Willis Carrier installed the first commercial air conditioner in a Brooklyn printing plant that consumer cooling products became available. By the 1920's residential refrigerators were introduced, but the non-electric "ice box" was not retired in most American homes until the early 1950's. Recent federal energy regulations have caused a significant improvement in the efficiencies of standard residential refrigerators built after 1993; however, even these energy efficient models can still represent over 20% of your monthly electric bill.

If you plan to stay on the utility grid but want to minimize the electric loads on an emergency generator, consider replacing your old refrigerator with a new energy efficient model. Another solution is to purchase one of the new super efficient top loading freezers which can be separately wired to the emergency backup system. As long as the freezer is not constantly opened, it will require much less compressor run time and will keep food frozen longer with power off than a standard refrigerator freezer compartment.

If you are planning a weekend cabin and your refrigeration needs are minimum, there are several portable chest refrigerators designed to plug into a car cigarette lighter socket. These units use a solid state device to produce the cooling effect without a motor driven compressor. Some are large enough to keep milk, soft drinks and sandwich meat cold. By purchasing a second connecting cable and removing the end plug, this unit could be direct wired to a 12 volt DC battery system or solar modules.

Those of you desiring a more permanent refrigeration system for an off-grid home should use either a propane powered refrigerator or a super efficient low voltage DC model designed for alternative energy applications. Unfortunately, neither type are frost proof and will require periodic removal of ice buildup, but this is much easier to do with today's smooth interior wall designs and non-metal materials that are easy to clean.

Figure 4 - 1 Modern 24 volt DC Sun Frost RF-16 refrigerator and freezer.

When your home will be unoccupied for long periods of time, the 12 or 24 volt DC models would not require a 120 volt AC inverter or gas flame to operate, and the low energy drain could be offset by a small solar array since all other electrical loads would be off.

If your system will not have a solar array but will use a backup generator, a propane powered refrigerator would be ideal if the generator and cooking stove also operated on propane. A common propane storage tank and gas manifold could be used to supply all gas appliances.

Figure 4 -1 shows how this author's high efficiency 24 volt DC Sunfrost refrigerator can be easily integrated into conventional kitchen cabinets and counters.

The following chart lists several of the most common sizes of very energy efficient refrigerators and freezers, and their average electrical or propane energy consumption.

TABLE 4 - 1
SUPER EFFICIENT REFRIGERATORS and FREEZERS

Manufacturer	Model #	Capacity	Dimensions	Energy Usage
Sunfrost	F-10	10 ft.3 Freezer only	43 1/2” high 34 1/2” wide 27 1/2” deep	55 amp hours per 24 hours 12 volt or 24 volt DC
Sunfrost	RF-12	12 ft. 3 Freezer only	49 1/2” high 34 1/2” wide 27 1/2” deep	28 amp hours per 24 hours 12 volt or 24 volt DC
Sunfrost	RF-16	16 ft.3 Refrigerator Freezer	62 1/2” high 34 1/2” wide 27 1/2” deep	45 amp hours per 24 hours 12 volt or 24 volt DC
Sunfrost	RF-19	19 ft. 3 Refrigerator Freezer	64” high 34 1/2” wide 27 1/2” deep	62 amp hours per 24 hours 12 volt or 24 volt DC
Norcold		19 ft. 3 Refrigerator Freezer	61 1/2” high 25 1/4” wide 23 3/4” deep	Propane fuel 0.39 gal per 24 hours (1.67 lbs./24 hours)
Vestfrost	SKF375	10.4 ft. 3 Refrigerator Freezer	78 3/4” high 23 3/8” wide 23 3/8” deep	0.70 kWh per 24 hours 115 volt AC only
Vestfrost (Top Load)	SE215	7.3 ft. 3 Freezer only	33 1/2” high 44 3/8” long 26 5/8” deep	0.54 kWh per 24 hours 115 volt AC only
Servel	RA1302	7.7 ft. 3 Refrigeration Freezer	58” high 25” wide 25” deep	Propane fuel 0.40 gallons per 24 hours
Novakool (RV & Marine)	R3800	3.5 ft. 3 Refrigeration Freezer	28 3/4” high 20 14/” wide 18” deep For Built-in	0.48 kWh per 24 hours 12 volt or 24 volt DC
Frostek (Top Load)		8.5 ft. 3 Freezer	38” high 44” long 31” deep	Propane fuel 0.36 gallon per 24 hours

The refrigerator is one of the few major appliances in your home that is almost always consuming electricity, day and night, winter and summer, and is one of the main appliances you will want to keep operating during a power outage. This is why selecting your next refrigerator can be a major step in your energy independence.

If the super-efficient 12 and 24 volt DC refrigerators and the non-electric propane units listed in Table 4-1 do not fit your needs or budget, there are several U.S. manufacturers starting to build very efficient 120 volt AC refrigerators and freezers. These will still require an inverter or generator to operate when the grid is down, but their lower power requirements and thicker insulation will provide much better emergency performance.

All of the following models have automatic defrost which requires higher energy input than manual defrost models, but these are still much more efficient than lower cost standard efficiency models. Adding an automatic ice maker option will add additional energy consumption to the figures given.

In the 14 cubic foot range, the CTL and CTN series by Magic Chief with top freezer door, have an annual energy budget of 437 kWh per year. In the 18 cubic foot range, the JTB and MTB series by Maytag with top freezer door, have an annual energy budget of 485 kWh per year.

In the larger 21 cubic foot range and freezer door on the bottom, The BX and BP series Amana and the 596 series Kenmore both have an annual energy budget in the 594 kWh per year range.

For a rough comparison, the highest efficiency rated 14 cubic foot Sunfrost refrigerator freezer requires only 339 kWh per year to operate.

When shopping for a new refrigerator, be sure to compare the posted annual energy tag on the unit you are considering with the above very efficient models. Remember, the same manufacturer will also have models with much higher energy usage to reduce manufacturing costs so be careful and read the small print!

CHAPTER V
WATER, WATER, EVERYWHERE ....

Can you pump water with an alternative energy system? What type of well works best, do you need a storage tank, do I need special filters? Outside of lighting, pumping water is one of the main reasons remote sites need a solar photovoltaic or generater system. Since this may also be the largest individual electrical load on the system, and a well may be drilled before installing the power system, we need to be sure things are done correctly in the beginning.

WELL PUMPS: Let's start with basic well information. The majority of homes not served by a public water service utilize a drilled or dug well. Unfortunately, many of these wells are fairly deep which require larger well pumps and 240 volt electrical power to keep down the electric current required. The higher voltage allows using a smaller wire size for what may be the longest wire run in the system. Since most alternative energy powered homes use 120 volt AC inverters, the most immediate problem is how do you operate a 240 volt AC pump wired with a cable having three conductors and a ground, from a 120 volt AC inverter or generator having a cable with two power wires and a ground.

If you are still in the preliminary design stages and the well has not been installed, request the well driller to install the 120 volt version of the well pump he normally furnishes, and request the pump motor to be a low in rush current design with the highest efficiency available. Since the lower voltage motor will require double the current rating on the wire supplying power, this will increase wire size and cost. In addition, this very long run of wire will have much higher voltage losses if undersized, which will reduce pump performance. Do not undersize this well pump wire thinking you are saving money.

If the well is existing, and changing the well pump and wiring for a more energy efficient model is not in your budget, you can operate the pump directly from a

120/240 volt generator. Since the pump will not operate when the generator is off, many systems utilizing this design include a water storage tank, with the generator powered pump filling up the tank during the several hours per day the generator is operating.

When the generator is off, a smaller 120 volt AC or low voltage DC pump can be used to pump water from the storage tank to the pressurized domestic water distribution system from the house. Although the storage tank has many advantages, extra care is required in the design and installation of this type system due to the higher risk of freezing and water contamination. For those systems with an existing 240 volt well pump that cannot be replaced and must operate when a generator is not running, use a 120/240 volt step up transformer. They are reliable, have no electronics to fail, but can cost almost as much as changing to the lower voltage, higher efficiency pump.

EXPANSION TANKS: The expansion tank is one of the most misunderstood devices in a plumbing system and is usually never sized correctly for energy efficiency. Early models consisted of an inverted sealed steel tank having an air pocket "trapped" in the top, with the water entering and exiting from the bottom. Water cannot be compressed to build up pressure like you can pump air in a tire to raise tire pressure, so using trapped air in a water system allows compression which pressurizes the system.

Over time the trapped air in the earlier expansion tanks would start to dissolve into the water and eventually the tank will completely fill with water. This was called "water log" and once the air was no longer in the tank to compress against the water, the system could not "store up" system pressure. Once this happened, every time someone opens a faucet the pressure switch would sense this instant drop in system pressure and the pump would start. As soon as the faucet was closed, the pump would immediately stop as the pressure switch would instantly reach set point since the system had lost all expansion capacity.

Most expansion tanks now consist of an inverted steel tank that is pressurized at the factory. Inside is a rubber bladder connected to the bottom inlet. As water enters this bladder connection, the bladder begins to fill just like a balloon, until the air pressure inside the tank equals the pressure of the water inside the bladder. In most homes this is adjusted to a pressure of 40 psi, although multi-story homes may require a 50 psi cut-off pressure to maintain acceptable shower pressure on the upper floors. For each 2.31 foot of height a faucet on shower is raised above the pump, an additional 1.0 psi of pumping pressure is required to maintain the same water flow. See example on page 170.

EXPANSION TANK SIZING: Since expansion tanks take up space in a conventional home and cost more in larger sizes, most home builders install the smallest tank possible. The smaller expansion tank will still operate satisfactorily, but the smaller storage capacity requires the pump to start and stop more frequently. For example, most commodes require three gallons of water for each flush. If the expansion tank only has a two gallon storage capacity, the well pump must turn on and off every time the commode is flushed.

Figure 5 - 1 Large pressurized well expansion tank and dual cartridge filters.

Since many showers have a 2 to 3 gallon per minute flow rate, using a shower with a small expansion tank may require the well pump to turn on and off every two minutes during a shower since the pumping flow will greatly exceed the shower flow and quickly refill the small expansion tank.

We recommend purchasing the largest expansion tank you can buy and still have room to install. They are relatively inexpensive, but can take up as much room as an apartment sized refrigerator. A large expansion tank allows multiple commode flushing and longer showers before the pressure drops enough to start the pump. This means much longer "off" time for the pump and longer "on" time when it does operate. This greatly reduces wear and tear on the pump and inverter or generator supplying the electrical power. Figure 5 - 1 shows a large size expansion tank and the space required for its installation.

When you turn on a light bulb, the electric current flow when the light is initially turned on is almost the same as it is after it has been on for several hours. However, when you start a pump or other motor driven load, the initial current flow, or "in-rush", can be three to four times higher than the current flow after the motor is started! This is obviously very abusive to the start-stop controls, motor windings, and the inverter or generator supplying the power to the motor. A momentary large current in-rush may also cause the circuit breaker on an inverter or generator to trip, even though the normal operating current and voltage of the pump is within normal operating limits. Anything you can do to reduce how often a motor driven load is started, will greatly improve system performance and reliability.

WATER FILTERS: Today's water conservation demands for reduced flow shower heads and toilets has forced manufacturers to use smaller flow valves and orifice holes. Although any well supplied residential water system will benefit from a replaceable cartridge water filter, the new low flow plumbing fixtures demand better filtering to reduce sand particles picked up by the well pump from blocking the smaller orifices. Replacement of these filters every three months also reduces pumping and electrical loads as the filters begin to "fill up."

DOMESTIC WATER STORAGE TANKS: One of the most difficult electric loads to supply with an alternative energy system is a deep well pump. The combination of a large electric motor with short run times and multiple start/stop cycles produces heavy electric demands on any inverter and battery bank.

Earlier in this Chapter I discussed how these multiple start/stop cycles can be reduced by installing an over sized expansion tank. For most projects this will provide an adequate solution; however, there are many locations having very deep water tables or very low flow rates that will require either a very high horsepower deep well pump or a very slow pump respectively. For these special cases a non-pressurized water storage tank and second pump can solve both problems.

Figure 5 - 2 shows a commercially available 500 gallon un-pressurized plastic tank designed for domestic water storage. The tank will hold several days of water usage for a typical single family residence.

Figure 5 - 2 500 gallon water storage tank and 24 volt DC booster pump and in line filter.

The tank is filled by a deep well pump which only operates when a generator is operating. A ball float switch hanging inside the tank allows the deep well pump to run when the tank is low and turns it off at the high level point.

Since the tank is not pressurized, a second pump is used to draw water from the bottom of this tank and into the home's pressurized plumbing system. This system can also be used to match a very low flow well to a residential plumbing system.

Many wells have a very slow refill rate and cannot support a high flow pump even if it only operates a few hours per day. For these applications, or off-grid power systems not capable of powering a large pump, a very small low flow pump is installed in the well or creek to slowly refill the tank. The small pump may need to operate many hours per day of low water flow to match the shorter periods of high pressure flow rates out of the tank, but the smaller pump more closely matches the low flow well or limited capacity electric system.

BACTERIA FILTERS: Most conventional deep wells supplying a residential water system do not involve an open storage tank and any possible introduction of bacteria is limited to what enters through the hundreds of feet of porous rock and sand of the earth's surface. However, some remote water systems must utilize water pumped from a pond or stream when deep wells are not practical, or open storage tanks are needed to assist a low flow deep well.

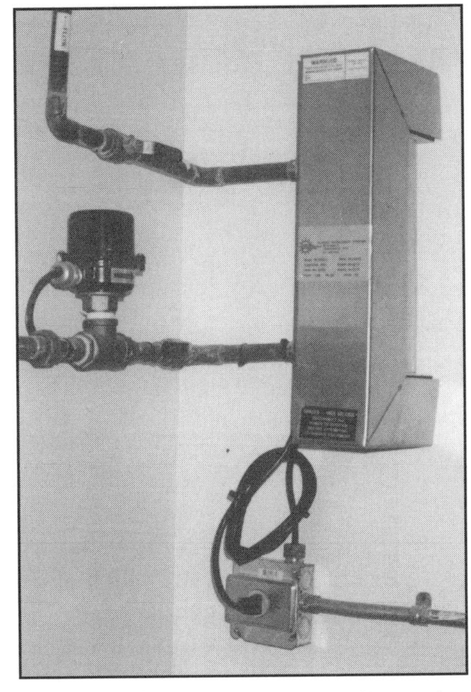

Figure 5 - 3 Ultraviolet bacteria filter and flow switch system.

These applications have a much higher risk of bacteria contaminating the water supply and an in-line ultraviolet water purifier down stream from the sand filter or open storage tank is strongly recommended. These units have no moving parts to wear out and only require a low wattage ultraviolet fluorescent lamp to function.

These filters are actually a pressurized chamber surrounding a clear glass tube around the ultraviolet fluorescent lamp. As water flows through the chamber, all bacteria travels along the length of the fluorescent lamp and are destroyed. Ultraviolet light in the 254 nano-meter wave length of proper intensity is very effective in killing bacteria, mold spores, protozoa, virus, and pathogenic microorganisms. The piping arrangement for two replaceable carbon filters with bypass line shown behind the storage tank in Figure 5 - 1, water to an ultraviolet bacteria filter and in-line flow switch as shown in Figure 5 - 3. Note the in line flow switch used to activate the filter.

A typical 7 gallon per minute capacity ultraviolet water purifier will consume 70 watts of electrical power and this should be enough flow capacity for most three bedroom homes. It must operate any time water passes through it, as any bacterial laden water allowed to pass through when it is momentarily turned off can contaminate your entire plumbing system. Many installations include a fast acting flow switch to turn on the unit when the water flow starts. This keeps any untreated water from passing the fluorescent lamp before it is turned on and reduces standby electrical usage; however, some forms of bacteria can "swim" several feet in still water which could allow them to pass through the filter while it is off due to no flow. If your water system requires a very high level of protection, it may be safer to allow the purifier to operate continuously and live with the penalty of shorter lamp life and higher energy usage.

CHAPTER VI
UNDERSTANDING THE BASICS

Before going further, a brief review of the electrical terms we will be using may be helpful. When describing the electric grid, we compared electrical flow in wires to the flow of water in a pipe. At this point there is no need to get into current and voltage phase, transformer effects, or voltage wave forms - so let's keep it simple.

When we turn on a light, electricity "flows" through the lamp filament causing it to heat up and glow, which produces visible light and radiant heat. This flow of electrical energy, which is measured in amps, is very much like the flow of water in our plumbing system, which we measure in gallons. When we introduce the factor of time into this measurement, we are defining work or energy. If we can fill a gallon bucket from a ½" pipe in one minute, our pipe is providing a flow of one gallon per minute.

The lamp cord in our example could be carrying an electrical flow of one amp to power the light. If the light operates for one hour, it has consumed one amp-hour of energy. An 800 amp-hour battery that has stored up enough energy to deliver 800 amps of electrical energy for one hour, also could deliver 400 amps for 2 hours, 100 amps for eight hours, or 40 amps for 20 hours. The 20 hour rate is the standard time period for measuring battery capacity. A battery's capacity to deliver power is not linear, and as the electric load is increased, the discharge time will rapidly decrease resulting in less total energy than would be available over a longer time period.

Going back to our plumbing example, if we wanted to increase the flow of water from one gallon to two gallons per minute, we could increase the size of the pipe to allow more water to flow. Although this is easily understood, it is not possible with an existing plumbing system. When we need to increase water flow and the pipe size cannot be changed, we can increase the water pressure. As the pumping pressure is increased, more gallons of water will flow. A continued increase in water pressure will increase flow until the maximum flow capacity of the pipe is reached. Any pressure increase beyond this point could cause objectionable noise and vibration, or even a rupture in the pipe.

The energy flow rates in electric circuits, measured in amps or amp-hours when time is being considered, is also affected by the pressure in the electric circuit. This electrical pressure is called "voltage" and the higher the voltage, the more pressure there is to move the electric current through the wiring. As in the plumbing example, we can increase the flow of energy through a given wire size by increasing the circuit voltage. This energy flow can be increased beyond the design capacity of the wire to the point the wire melts from the heating effect of the high current flowing through the resistance of the small wire size.

We can measure voltage in a circuit but that only tells us the "pressure" of the current flow, not the quantity of energy available. We can measure the current flow in a circuit but that also does not tell us how much flow occurred in a given time, or the capacity of this flow to do useful work. Since we are really interested in energy consumption, not energy potential, we need a measurement of energy or work provided by the electrical circuit for the load. This energy consumption is measured in "watts". By definition, one watt equals one amp of current flow driven by one volt of circuit pressure.

$$\text{WATT} = \text{VOLT} \times \text{AMP}$$

A single 100 watt light bulb will allow 0.83 amps of current to flow through the filament when connected to a standard 120 volt AC wall outlet.

$$100 = 120 \times \text{AMPS}$$

$$\text{AMPS} = \frac{100}{120} = 0.83$$

If we continue to operate this lamp for ten hours, it will consume 1,000 watt hours of energy.

Energy Consumed = WATTS x HOURS

(100 watts) (10 hours) = 1,000 watt- hours

The unit of 1,000 watt-hours is commonly referred to as one kilowatt hour (1 kWh). Depending on your location, electric utility rates will probably average between 6¢ to 12¢ per kWh in the United States. In Europe, rates of 15¢ to 30¢ per kWh are not uncommon. Localities served by inexpensive hydro power may be lower, while highly populated areas served by oil or gas fired plants to reduce coal fired emissions may be a higher rate.

UTILITY kWh METERS: The utility meter on your house is designed to measure the combination of voltage, amp flow, and time and record kilowatt hour consumption (kWh). In most cases these meters are not reset to zero, so this month's reading is subtracted from last month's reading, and the difference is multiplied by the cost per kilowatt hour and the bill is then calculated. Like everything else however, it's not quite that simple. Look at your most recent electric bill. Depending on the utility company, you will see many additional charges that were added which had nothing to do with the total energy you consumed. For example, the bill will usually start with a meter base charge which is a fixed cost each month just to have the power available to you, whether you use it or not.

DEMAND CHARGES: Some residential rates and almost all large commercial rates also have a kW demand charge or a time of day penalty charge added. This additional fee or "fine", is a penalty being charged during a given set of hours each day when the demand on the electrical grid is highest. The grid system and generator plants must have the capacity available to meet these peak loads at all times; however, these daytime loads will be much lower during the evening.

This loss of revenue during low demand periods does not pay for the added equipment capacity when it is not being used, so the time of day penalty charges are imposed to recover the added standby or idle costs. Although this may seem unfair to the end user, most utility companies would rather have a constant electrical load even if this means a lower total to sell, since idle equipment capacity costs can be reduced. For this reason, many utility companies are now offering very low off peak rates to customers who can shift their electrical loads to the evening hours when demand is lower. In addition, the investment cost for new generating plants to operate during the peak periods can be avoided.

You may still want to be connected to the utility grid and only use an alternative energy system for emergencies. If you are on a time of day rate, you could significantly reduce your electric bill with the same emergency backup system. By charging a battery bank and operating all heavy appliances at night during relatively low electric rates, then switching over to the emergency battery system during the higher cost daytime peak hours, it is possible to save up to 30% of your monthly electric charges. This strategy would require not operating unnecessary appliances, air conditioning, or large motor loads during these peak daytime hours.

Almost all modern alternative energy and battery backup power systems require automatic transfer switching and battery charging features, and many also include programmable clock control features that can provide this switching automatically. What was envisioned as an emergency backup power system only, could actually pay for itself if your utility company offers this time of day rate schedule. Although emergency generators can and often are operated at many large hospitals and institutional plants during summer afternoon peak periods to reduce their electrical demand penalty charges, this has not always produced the savings expected.

Regardless of fuel and generator type, you can never generate your own electricity as cheaply as a large power plant. Initial utility bill savings can become insignificant when faced with the cost of a major generator overhaul or reduced emergency standby reliability as the wear and tear of daily operations begins to take its toll on the equipment. Determining whether load shifting or peak shaving is cost effective requires a detailed analysis of your electrical loads, and the time of day these loads occur.

APPLIANCE EFFICIENCY EFFECTS: Suppose we want to power a critical load from a battery bank that is recharged by a solar array. Without getting into efficiency losses of the charging cycle, or the weather effects on solar array performance, let's take a very simple example. A 75 watt incandescent table lamp must operate 12 hours per day. Although solar photovoltaic system costs are declining, in smaller system sizes we will probably pay about $300 for a 40 watt solar module ($7 to $8 per watt). In most states, a fixed mount solar module will receive full sun from 9 AM to 3 PM, or approximately six hours on a clear day.

Energy Usage	=	Energy Collection
(75 watts x 12 hours)		(40 watts x 6 hours per day x 4 panels)
= 840 watt-hour		= 960 watt-hour*

*Note that the 960 watt-hour value is a peak capacity of the solar modules and the daytime average generation will probably be closer to our actual usage of 840 watt-hours.

From the above example we find that it takes almost four solar modules to collect enough energy in six hours, to operate a 75 watt bulb for twelve hours. Using an average cost of $300 per module, this solar powered energy system will cost over $1,200 (4 panels x $300) just to power a 75 watt light bulb. Since this is obviously a very expensive lighting system, let's reconsider this analysis by substituting a 26 watt compact fluorescent lamp which has the same lumen light output as the 75 watt incandescent bulb. We have just reduced our lighting load by 2/3 and only one solar module will now be required to provide our lighting energy needs. By substituting a $10 compact fluorescent lamp for a $1 light bulb, we saved almost $900 in solar array cost, with no loss in light output or operating hours. A similar cost savings will also be realized in battery costs, by requiring one-third the original system battery size.

When comparing the wattage values of energy efficient appliances given in the Appendix against the nameplate data for your existing appliances, notice how more efficient appliances or "Energy Star" components can save more in energy usage costs than their additional purchase cost. Although not as critical for emergency backup systems which are only required to operate a few hours per year, low efficiency appliances and lights have no place in off-grid independent energy systems. With the high cost for alternative energy components, only the highest efficiency lighting and appliances should be used and all low efficiency equipment should be replaced, or only operated when utility grid power is available.

ELECTRIC MOTORS: A brief review of electric motors will be helpful along with understanding the differences between AC and DC motors. Since many appliances operating on an alternative energy power system have motors, the type, voltage, and efficiency ratings of these motors can affect your system's overall efficiency. As soon as the new air handling unit was delivered to my home during construction, I removed and discarded the standard efficiency fan motor, and replaced it with a premium efficiency motor. This may seem extreme, but for motors required to operate many hours per year, it can be cost justified.

AC MOTOR OPERATION: Almost all AC powered motors are "squirrel cage" design. Without getting into motor theory, if we took an AC motor apart we would find a stationary coil of wire in the motor housing directly connected to the power wiring feed. The rotating motor shaft contains a second coil of wire, but this rotating coil is "self-excited" and has no wiring connection to the other coil or stationary parts. The electricity flowing in this rotating coil is generated by the nearby stationary coil around it, which is constantly reversing its magnetic field as the alternating AC electricity flow changes direction 60 times each second (60 cycle in the United States, 50 cycle in Europe). AC motors now can be purchased with a "premium efficiency" rating and AC motors required to start heavy loads can include a capacitor which significantly reduces high in-rush starting currents.

DC MOTOR OPERATION: Unlike an alternating current (AC) motor, a direct current (DC) motor is connected to a constant voltage supply that does not reverse flow or rapidly change voltage level. A constant voltage flow in a DC motor coil does not generate electric flows in the nearby rotating coil, which means all DC motor designs require sliding electrical contacts to transfer electrical energy to the rotating coil.

To reduce wear, carbon brushes are used which rub against copper contacts around the motor shaft. Unlike AC motors, most DC motors have brushes that wear out, and these brushes sometime "stick" or build up deposits on the copper shaft contacts which must be cleaned. It would not be unusual for a low cost DC motor to need brush replacement or disassembly and cleaning every two to three years, and this added motor maintenance may be beyond the skill level and patience of a homeowner expecting the DC appliance to always operate when it is turned on.

With today's high efficiency inverters, most people prefer to use AC motor driven appliances and allow the inverter to convert the DC battery power into AC power. This slight efficiency penalty is usually acceptable to avoid the higher cost and brush maintenance required for a DC motor.

CHAPTER VII
BACKUP POWER SYSTEM TYPES

Flashlights: Probably the most basic backup power system you can own is a flashlight. It provides light when the utility grid is off, it is self contained with its own control switch and battery, and many can be recharged and left plugged in while waiting for the next power outage. Although flashlights are extremely inexpensive, it is amazing how many homes and apartments are without one, or they cannot be found when an emergency occurs. Newer style lanterns with fluorescent lamps and rechargeable batteries can provide room filling light with only a fraction of the battery drain of older incandescent flashlights. Buy several and know where they are.

PORTABLE GENERATORS: The next most common a backup or emergency power system is a portable generator. These are most effective when you have advance notice of a potential storm related power outage. Having advance notice allows time to get the generator refueled and warmed up. There is nothing more difficult than trying to fuel and start a generator with an empty gas tank, that hasn't been started for six months, during freezing temperatures, in the dark.

Unlike a battery powered system, a generator will not run out of power, (as long as it does not run out of fuel). Fairly small generators can power large electrical loads for long periods of time, but they are also noisy, produce deadly exhaust fumes if operated in confined spaces, and require the handling of flammable fuels around hot exhaust surfaces during refueling. Generators are a great choice in rural areas to keep a well pump or freezer operating along with a few lights. Smaller portable generators are not designed to operate continuously, and when they run out of fuel, the lights go out.

BATTERY POWER: Battery powered homes are not new. Until the 1930's many parts of western farm and ranch lands were without utility power. Low voltage DC lighting and kitchen appliances were connected to a battery bank. These batteries in turn were recharged by a windmill driven DC generator. Beginning with the windmill driven pumps of the 1860's, these early designs were later adapted to drive electric generators. There were over 8 million windmills installed before the western rural electrification program took over in the 1930's. These wind charger or "Delco" systems provided reliable electricity to western farms still years away from the slowly advancing utility grid.

These systems had limited capacity, but low voltage DC lights, radios, and small motors from heater fans were easily obtained from junked cars. Even though a battery based power system introduces the need for system and battery maintenance that most home owners are not prepared for, most farmers were already maintaining far more complex farm machinery, and the required tools and spare parts were always nearby.

Figure 7 - 1 Still in operation - 100 year old western windmill.

Today it is still possible to install a low voltage DC battery backup power system, but the choice of lighting and appliances will be limited to the 12 volt equipment manufactured for the recreational vehicle and boating industry. This may be an ideal solution for a small cabin, but the 12 volt DC low voltage wiring requires much larger wire sizes and heavier switches than a 120 volt AC system. For example, each 60 watt light bulb on a 120 volt AC system will draw ½ amp.

$$\text{WATT} = \text{AMP} \times \text{VOLT}$$

$$\frac{60}{120} = 0.5 \text{ amp}$$

A 60 watt bulb designed to operate on a 12 volt DC system will draw 5 amps, which is ten times the 120 volt current!

$$\text{WATT} = \text{AMP} \times \text{VOLT}$$

$$\frac{60}{12} = 5.0 \text{ amp}$$

Most residential 120 Volt AC lighting circuits are wired using #14 wire which requires a 15 amp circuit breaker to limit the current to the maximum this wire size can carry safely. Using a 12 volt battery for power, three 60 watt 12 volt light bulbs would exceed this circuit breaker rating. There also seems to be a myth that low voltage systems are safer. Although we do not hear of anyone being electrocuted by a 12 volt car battery, if you have ever made the mistake of touching a screwdriver to the positive and negative terminals of a car battery at the same time, you will find melted metal and the remaining metal surfaces too hot to handle. This would easily ignite adjoining building materials and could cause a fire if the circuit wiring is over loaded, regardless of the relatively safe low voltage. In addition to the higher cost of larger wire sizes, there are also other problems associated with a low voltage battery system that make larger systems impractical including the need for heavier duty switches and circuit breakers.

INVERTER POWER: Chapter IX will go into more detail on inverter based backup power systems. This system converts the stored DC electrical energy from batteries into 120 volt AC to allow the use of standard 120 volt AC appliances and lights without operating a generator. The battery size and system loads determine how long the inverter can operate without having to recharge the battery bank.

SOLAR POWER: By adding a solar photovoltaic array and charge controller to any battery bank, you can increase the time the system can supply power before needing recharging by the grid or a generator solar systems will be discussed in much more detail in Chapter XI. A large solar array may totally eliminate the need for a generator, but high costs usually prohibit solar designs that generate more than two to three days of excess energy storage during cloudy weather.

WIND POWER: We all know the wind blows where we live, and sometimes with a very destructive force. It is not the occasional gusting winds that creates the potential for powering your home from a wind driven generator, it is the lower but constant wind which creates the best opportunity for "free" electricity. The fact is, some parts of this country and most coastlines have fairly constant winds and other areas do not, and there is not much we can do about that. Of course the higher we go above the earth's surface, the stronger and more constant the wind patterns are, and a high tower can provide additional wind energy for a surface location that has only minimum surface winds. Wind turbines on tall towers, however, are not for the faint of heart as a tall tower greatly increases periodic maintenance and repair costs, and the potential for property liability from a catastrophic structural failure or lighting damage.

Several low maintenance wind driven generators are now available which are designed for mounting above the ridge line of a roof. Two and three bladed models having rotor diameters up to six feet are still small and lightweight enough for a single pole mount. Almost all turbine designs include either twisting blades or tilting mounts to reduce or stop rotation during destructive wind gusts. In normal operation, these smaller turbines will start generating power when the wind exceeds 7 mph, and can provide 200 to 800 watts continuous output in 15 to 35 mph winds depending on blade diameter and generator size. Although a single wind turbine may not provide all of your power needs, there are many times the wind blows after sunset and during cloudy winter weather which makes it an ideal supplemental power source for a solar installation. Since the wind turbine can charge the same battery bank, the only added cost is a second charge controller.

Some of the lower cost and smaller diameter wind generators may operate in excess of 2,500 RPM in high winds which can produce a very objectionable noise. Care must be taken when selecting the ideal mounting location and stay as far away as possible from all structures, chimneys, or vents that can disrupt natural air flow patterns. Most system designers recommend a minimum mounting height of 30 feet.

CHAPTER VIII
BATTERIES AND BATTERY CHARGING

We have briefly referred to using and charging batteries with solar photovoltaic arrays and generators, but there are many types and sizes of storage batteries, having different levels of maintenance requirements and reliability. To reduce automobile weight and cost, and increase fuel economy, car batteries are made with very thin lead plates. Automobile batteries will quickly fail if subjected to the heavy current flows and daily deep discharge/recharge cycles found in most battery based power systems. Automobile batteries are only designed for a very short period of heavy load and then a long period of recharging.

Golf cart batteries are available locally and their heavy lead plates are designed for a daily deep discharge cycle. The 6 volt golf cart battery is the minimum acceptable battery for an alternative energy power system. They are relatively inexpensive, can be totally recycled when replaced, and should last three to five years under normal use. Many residential battery based inverter systems start out using lower cost golf cart batteries, and after reaching their useful life are replaced with the heavier L-16 electric vehicle battery. These batteries are similar in footprint size, but taller and heavier. The L-16 series battery can provide up to eight years of reliable service if not abused.

Industrial tray batteries are extremely heavy and the smallest still may weigh over 1,000 pounds, but their tall cells are designed to last over ten years. Individuals planning off-grid homes in very remote areas where periodic battery replacement is impossible should consider this more expensive but longer life battery. A note of caution is in order about battery selection. Most deep discharge batteries are designed for regular daily or weekly discharge cycles throughout their useful life. However, if you are planning to install a battery backup system that operates only during an occasional emergency power outage, these once or twice per year discharge cycles will cause these batteries to buildup a plate resistance to charging and can reduce their useful life to less than four years. For occasional use only applications, you may want to consider a 2 volt glass cell battery. These batteries are expensive, heavy, and will spill acid if tipped, but they are almost indestructible. These batteries are designed for occasional use emergency uninterruptible power systems for the computer and communication industry, and will last over 25 years if properly maintained.

BATTERY TYPES: Whether you are planning an emergency backup power system, or a total alternative energy system, you will probably be using lead acid batteries. Although the increased use of electric vehicles is driving new research in battery technology, most of these efforts are to minimize battery weight through the use of more expensive and exotic materials. Since weight is less a consideration in stationary battery systems, it is still difficult to find a more cost effective backup battery then the old reliable lead acid battery.

In an effort to allow using batteries in confined spaces and with minimum risk of acid spill and out-gassing, the lead acid battery technology is now available with a sealed battery case having either a liquid or gel electrolyte. The sealed case liquid electrolyte battery includes a special expansion chamber under the cell cap designed to recombine the venting gasses back into liquid form. A 12 volt deep cycle marine battery is typical of this design. The gel battery uses an electrolyte in a paste form and can be used in any position without fear of spills or hazardous gas discharge, but the gel battery still has about the same capacity and useful life as the sealed liquid electrolyte version which costs substantially less.

As long as the homeowner checks the water level in the battery on a quarterly or monthly basis depending on system demand, the advantages of a sealed battery can be matched with the much less expensive liquid acid batteries. If you plan to use gel batteries for your system, remember that the charging voltage is different from that used for a standard lead acid battery, so be sure the charger you use is designed to charge a gel cell battery. Unlike liquid electrolyte batteries, gel batteries do not require an equalizer charge.

BATTERY SIZES: The T-105 battery commonly referred to as a golf cart battery has three 2 volt cells with a nominal 217 amp hour rating at 6 volts and can store approximately 1 kWh of useful energy after system efficiency losses.

(217 amp/hr.)(6 volt) = 1302 watt/hr. Or 1.3 kWh storage capacity

Figure 8 - 1 shows a typical battery bank using sealed gel cell 6 volt deep cycle batteries mounted in a metal cabinet. Note that the use of gel cells, which do not have fill caps and do not require any maintenance, allows locating the batteries closer together. Metal battery enclosures are not recommended for system voltages above 48 volts DC. The T-105 series battery is approximately 7-1/8" wide x 10-1/2" long x 11-3/4" tall and weighs 62 lbs., with an expected life of three to five years in normal use.

The L-16 is the next most commonly used battery for off-grid and backup power systems. Its three cells provide 360 amp hours of storage capacity at 6 volts and has a life expectancy of five to eight years. It is 7"wide x 11-3/4" long x 16" tall and weighs 128 lbs. Be careful in comparing this battery between manufacturers as some versions have as few as 295 or as much as 390 amp hours of storage capacity, while still having the same physical dimensions.

Figure 8 - 1 6 volt gel cells mounted in metal cabinet.

The industrial tray battery is ideal for installations that allow the use of lifts or fork trucks to install and remove. This battery is actually a welded steel case, filled with 2 volt industrial cells. The cells come in several standard amp-hour sizes and any voltage or amp hour capacity is easily provided by combining different quantities of individual cells in a custom made fabricated metal case having different lengths and widths.

BATTERY CHARGING: Battery charging up to now has been considered a single event, but in reality, is a multi-step process which requires careful control of the charging voltage, charging rate, and battery temperature. In addition, some charger manufacturers feel the charging voltage should be "pulsed" on and off while others rely on a constant current flow of gradually reduced intensity during the charging cycle.

Most of us have used a "trickle" charger to recharge a dead car battery. These low cost chargers consist of a transformer to reduce the 120 volt AC line voltage and a rectifier to convert the alternating AC current into a DC current flow. These units are very reliable for this type of occasional battery recharging, but are totally wrong for an inverter or generator based system since they have little or no voltage control which can over charge and "boil over" a battery left alone too long. When we say 120 volts AC grid power, the 120 volts is actually the average voltage of a current flow that is reversing 60 times each second. (50 cycles per second in Europe.) The instantaneous peak voltage of this average 120 volts reaches 169 volts.

The transformer and rectifier in a low cost trickle type battery charger use only this 169 volt peak of the line voltage to operate, with the balance of the voltage wave form supplying no charging energy at all. Although any AC generator can have the same instantaneous peak voltage of 169 volts, most generators are designed to drop the peak voltage as more loads are added in order to maintain their rated output. This can mean a low cost battery charger powered from a generator may not produce any charging output when the generator becomes loaded and the peak voltage falls.

When using a generator to operate a battery charger, you must remember that a battery charger having double the charge rate of a smaller charger will cut the generator run time in half; as long as it does not exceed the load capacity of the generator. Since most generators consume almost the same fuel rate at 75% load as they do at lower outputs, a larger battery charger will usually save generator fuel and generator run time.

When shopping for battery chargers, always choose the model that is designed to maintain its charge rate regardless of any voltage fluctuation in the electrical service it is powered from. These chargers are usually heavier, and advertised as suitable to operate from a generator. Many generators supply both 120 and 240 volts, but the total capacity is shared between two separate lower capacity circuits of 120 volts. Since most inverters and battery chargers operate from a 120 volt line voltage, look for the brand and model generator designed to maintain its peak voltage under varying load conditions, and with a switch to direct the full generator capacity into a single 120 volt service at full capacity.

All batteries will have a different rate of charge and discharge based on ambient space temperatures. Batteries are rated at 77° F, and a battery advertised to have 800 amps may only have a 400 amp discharge capacity in freezing temperatures, and 100 amp discharge capacity below zero. Inverters with built-in battery chargers and most industrial battery charges, have a temperature sensor built-in or remotely attached to the side of the battery bank. This temperature sensor controls the rate of charge which must be adjusted as the temperature of the battery changes. TABLE #5 in the Appendix provides the voltage and specific gravity for each level of discharge for the most common battery systems.

A heavily discharged industrial battery can take large current flows during the initial "bulk" charging period with little effect on long term battery life. A battery charger capable of these high current flows will recharge a battery much faster than a trickle charger. However, as the battery starts to reach its fully charged state, these high charging currents must be reduced during what is called the "absorption" stage to prevent large quantities of gas from forming as the water inside the battery is converted into hydrogen and oxygen. Hydrogen gas is extremely volatile and will explode if ignited by a spark in a confined unventilated space. In addition, constant hydrogen gas generation during each recharging cycle reduces the water level in the battery which requires constant refilling.

All high quality battery chargers include this voltage sensing control to vary the charge rate. As the battery starts to charge up, the battery charger should reduce the rate of charge. Once the battery is fully recharged, the charger should go into "float" mode in which only a very small charge is periodically applied to make up for any standby losses. This three-step, bulk-absorption-float, charging process will ensure long battery life and minimize battery service requirements.

BATTERY EQUALIZATION: A brief discussion about wet cell lead acid battery equalization is important as it is the only battery maintenance procedure that needs to be completed by the owner of a battery based system outside of cleaning and adding water. Remember, gel cell batteries do not require this battery equalization procedure.

Most battery storage systems will consist of multiple batteries and each individual battery will have multiple cells. There will always be some slight variations in battery plate material and acid electrolyte becomes more concentrated near the bottom of each cell after a large number of charging and discharging cycles which causes voltage differences from one battery cell to another. Since most battery chargers and photovoltaic solar charger controllers measure the battery voltage during the charging process to control the rate of charge, any irregular individual cell voltage can cause the battery charger to think all battery cells are at peak charge when most cells are not. When the total battery voltage indicates a fully charged state, several cells may still be below full charge which reduces the overall battery capacity. By deliberately over-charging a battery bank, referred to as "equalization," all under charged cells will be brought up to their full charge state.

During this equalizing charge process, any cells already at full charge will be over charged, resulting in the conversion of their liquid water and acid electrolyte into hydrogen and oxygen gas which will be vented out of the cells. Obviously, any time battery equalization is in process, there will be an above normal quantity of battery gas venting and a high water loss from the over charged cells.

Many battery manufacturers recommend going through an equalizing charge cycle whenever the differences in cell to cell voltage varies by more than 0.04 volts. Any battery system subject to a daily charge and discharge cycle should undergo battery equalizing at least once every three months.

Most high quality battery chargers and inverters that include built-in chargers usually have a switch for starting an equalization charging process. Be sure to follow the battery and charger manufacturers directions at all times when equalizing a battery bank.

There are several safety points that should also be addressed when starting an equalization cycle. Always increase the fresh air ventilation in the battery storage area to dilute the increased production of potentially explosive gases. Since you will be over charging the battery bank during equalization, a 12 volt battery bank may briefly exceed 15 volts and a 24 volt battery bank may briefly exceed 30 volts. It is recommended to turn off circuit breakers or remove fuses on all low voltage circuits during this process as the higher battery voltages may exceed the voltage rating of DC appliances and lights causing permanent over-voltage damage to these devices.

Battery equalization usually require four to eight hours of heavy battery charging to complete the equalization process depending on battery size and state of charge. This can be a significant load on a generator based system when utility grid power is not available. Equalization can also be accomplished with a solar photovoltaic array by increasing the setpoint of the voltage cutoff control on the photovoltaic charger if it does not include an equalization mode switch. This process should be started when the battery bank is already near a fully charged state, or there may not be enough hours of sun in a single day to complete the process.

PERIODIC BATTERY MAINTENANCE: All liquid electrolyte lead acid batteries form sulfate crystals which build up on the battery plates when the battery is kept below full charge for long periods or is rarely cycled. Over time, a poorly charged battery bank will have a much lower storage capacity since sulfate will build up and insulate the lead plates from the electrolyte.

Once this happens, the battery will permanently have a lower discharge capacity which is usually indicated by the battery heating up much hotter than normal during the recharging cycle. There are some chemical additives on the market that may slightly improve this condition, but good results are not always predictable or permanent.

Batteries will require occasional replacement of water lost during repeated charging cycles. It is rare that a battery will require adding water unless subjected to repeated extensive recharging cycles; however, the water levels of all cells should be checked monthly. Refill with distilled water only, which is available at most grocery and drug stores in gallon jugs for filling steam irons. You should also add water just prior to recharging which tends to cause circulation and mixing of the water and acid mixture.

At least twice per year or at signs of chemical buildup on the battery terminals, carefully brush water that includes a small box of dissolved baking soda over the batteries and terminals. Be careful to not touch any terminals during this procedure as water will conduct electricity even at these low voltages. Although not visible, there will be a film of acid that has settled on the top of the batteries which allows current to travel between battery terminals and drain some charge from the cells. Let this mixture neutralize the acid and dissolve away any chemical buildup on the terminals before rinsing with tap water.

Most owners of battery based power systems keep a battery log, like the example in TABLE 8 - 1, to record the condition of the battery over time. Select a representative sampling of cells that includes cells near main battery cable connections and cells in the center of each battery bank as it is not necessary to monitor every single cell. We are looking for trends or gradual changes in battery conditions since battery banks should be replaced as a unit, not one battery at a time. Record the specific gravity of the liquid electrolyte using a high quality battery hydrometer that provides a temperature corrected reading. Always take readings from the same cells, and be sure the battery is at rest for at least one hour prior to taking readings, and neither charging nor supplying any electrical loads at time of readings.

TABLE 8 - 1: Battery Log

Date	Battery Voltage	Room Temp.	Cell #1	Cell #2	Cell #3	Cell #4	Generator Meter Time

BATTERY LOCATION: The two most important considerations in battery location selection is ambient temperature and battery venting. The ideal battery temperature is 77°F, with 60°F to 90°F an acceptable range. As battery temperature drops below 60 degrees, the storage capacity begins to drop off. To avoid any reduction in battery performance, it is strongly recommended to locate the battery bank in a heated space having a concrete floor. The ideal heat source is baseboard hydronic radiation which does not create an ignition source or mix hydrogen gas with the return air like a ducted air heating system. Although most commercial systems include a separate room for the batteries like the one shown in Figure 8 - 2, most solar or backup power residential sized systems will utilize a battery box.

Due to the heavy weight of any battery bank and the potential for spillage, a concrete floor under the battery enclosure is almost a necessity; however, never place batteries directly on a concrete floor without an air space or insulation layer. The colder floor will cause uneven charging of cells due to the variation in temperature between the colder floor and the air, and any spillage will not drain away properly without an air space.

Figure 8 - 2 Separate battery room with raised battery bank.

An attached garage or utility room is an ideal location for a battery box, as these rooms are usually designed with floors to support heavy equipment and usually are not harmed by water spills. If these spaces are not conditioned, the battery box should be lined on the sides, top, and bottom with a hard foam type insulation board that resists chemicals and provides good thermal resistance.

A prefabricated fiberglass or aluminum tool box available for pickup trucks can make an excellent battery box for smaller systems; however, most battery boxes are fabricated on site using exterior grade plywood, 2" x 4" framing, and nail and glue construction. We have used the prefabricated all steel job site tool boxes, but found they will not last as long in the corrosive environment. All joints and wall penetrations should be calked and sealed, and the hinged access top should have a gasket. A minimum of two 2" to 3" PVC pipes should be used to vent hydrogen gas to the outdoors. These vent pipes should be located near the top of the enclosure and at opposite ends. An insulated wood 2" x 4" frame box with exterior dimensions of 3 feet wide by 4 feet long will hold twelve type L-16 batteries. A 3 foot wide by 8 foot long box will hold 24 batteries.

Battery banks have no moving parts, do not make noise, and after charging do not consume fuel to provide electrical power, but they do require taking common sense safety precautions and care in order for them to reach their rated life expectancy.

Figure 8 - 3 and 8 - 4 show a site built battery box large enough to hold 24 size L-16 industrial batteries. This box has been fabricated using cement and fiberglass matt applied over foam insulation board, a technique now being used to insulate commercial building exteriors.

Figure 8 - 3 Exterior view of battery box.

The insulated plywood top provides a very well sealed battery enclosure when the batteries must be located in unheated garages. Be sure to also insulate the bottom and use materials that can support the heavy batteries. Figure 8 - 4 shows how joints are sealed and all wiring enters through conduit that has been "fire stopped" to prevent hydrogen gas escaping into the room. A 2" PVC vent pipe is located at the top of each end, and a 2" drain pipe has been located at the bottom low point of one end. The drain should include a screw on cap to prevent a back flow of battery gas into the room when wind creates a pressure differential between the outside vent pipes.

To assist you in determining the physical space required for your battery bank, Table 8 of the Appendix provides typical wiring connections and overall dimensions for various orientations and quantities of 6 volt L-16 and T-105 golf cart batteries.

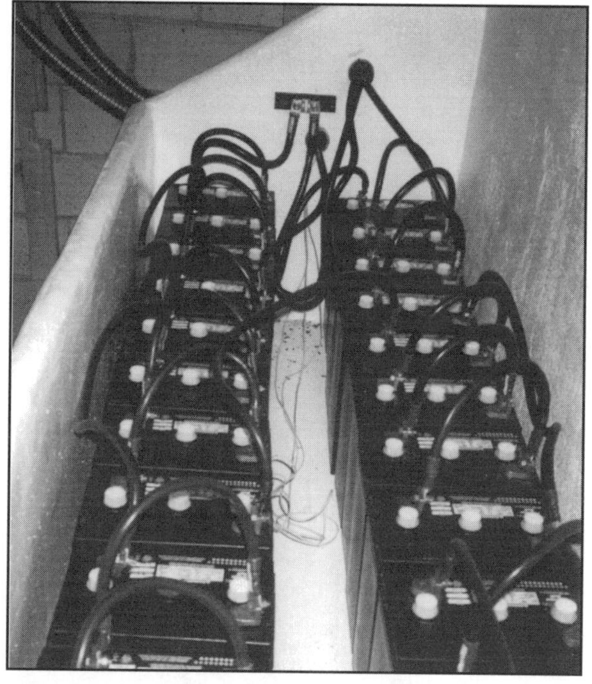

Figure 8 - 4 Interior view of battery box. Note catastrophic fuse to protect positive cable before exiting enclosure.

Use this Table to help you estimate the floor space your battery bank will require.

When selecting your battery bank, remember even new lead acid batteries have a 20% efficiency loss in the DC charging process, and there will be an additional 4% to 10% efficiency loss in the conversion of this energy back into AC electricity depending on the quality of the inverter. In other words, for every 100 watt hours of energy your solar system collects, you will only get back approximately 70 watt hours of useful AC power. Most system designers increase the size of the battery bank to take into account the efficiency losses in the charge/discharge cycle inherent in the use of battery based power systems.

CHAPTER IX
INVERTERS

We have discussed the pros and cons of AC generators and DC battery systems, but is there a happy middle ground that has the advantages of both? An inverter has heavy gauge wiring terminals on one side to connect directly to a battery or group of batteries. On the opposite side is a familiar duplex 120 volt AC wall outlet, or a connection point to supply a standard AC house circuit breaker panel. Between these wiring connection points is an electronic circuit that converts the high current and low voltage DC battery voltage into a lower current and higher voltage120 volt AC to power most household appliances and lights.

Until the electronic revolution, most inverters did not convert DC electricity into pure sine wave AC electric power, but instead generated a square or stepped voltage output that "approximated" the 60 cycle AC wave form. (50 cycle in Europe.) This was more than adequate to power most lights and motors, but could damage sensitive electronic components and produce a very objectionable hum in AM radios. However, these non sine wave inverters were relatively inexpensive.

Recent advancements in micro-electronics has greatly improved the performance of inverter systems. Almost all manufacturers now offer an inverter model capable of producing a very clean sine wave AC output like the Trace Engineering SW Series inverter shown in Figure 9 - 1, that may actually be more stable than utility grid power in areas where high electrical noise from other electrical equipment is a problem. These new sine wave inverters include extensive control features and built-in safety protection for high voltage, low voltage, high current draw, over temperature, and low battery voltage limit conditions. To stay competitive, many manufacturers still offer lower priced non sine wave inverter models for providing power to electrical loads that are not as sensitive to a non sine wave electrical source.

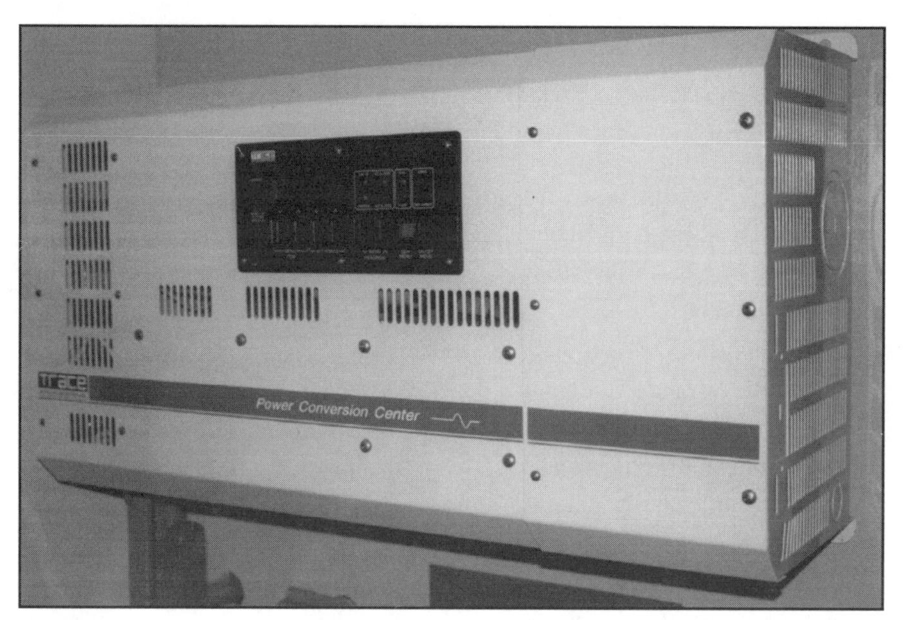

Figure 9 - 1 Trace Engineering 4,000 watt sine wave inverter with programmable battery charging and generator start controls.

Many inverters have electrical connections to direct wire a backup generator, with the inverter automatically starting the generator if the battery charge is low. The TRACE or HEART INTERFACE inverters are widely distributed and are recognized for their reliability in the residential and light commercial market. They are also packaged and sized to allow overnight airfreight shipping for factory repairs which make them ideal for rural sites having limited access to repair parts or service.

Using an inverter allows all house wiring, appliances, lighting, and controls to be conventional 120 volt AC equipment which is widely distributed and easier to install. Using a battery bank to power the inverter allows instant backup power when the utility grid fails and the generator is not used or not ready to start. Although not intended to serve as an uninterruptedly power supply for a computer, most new sine wave inverters are fast enough to sense a loss of power and switch from the utility grid to battery power before a desktop computer "crashes". How long the inverter can power your electrical loads depends on how large the load is, and how many batteries you have connected.

If you are only planning an emergency battery backup system for now, but are thinking about reducing your dependence on the utility grid in the future, this may be a good direction to take. Although purchasing an inverter larger than needed at this time is more expensive, you can always purchase more batteries and a generator at a later date to make the larger inverter more cost effective. You can also charge the same battery bank with a future solar photovoltaic array, wind generator, or water turbine, without having to replace the inverter. Most quality inverters include a built-in battery charger that will automatically recharge the battery bank when the utility grid is back on line or the generator is started.

Due to the recent increase in grid tied solar arrays that do not use battery storage, some inverters are now available to directly interface the DC solar array voltage to the AC utility grid. These inverters are usually designed for solar arrays above 48 volts DC and they do not include a battery charger mode or generator controls. Keep in mind that a grid tied inverter by design cannot provide emergency power from the solar array during a power outage.

WHAT INVERTERS WILL NOT DO: As described earlier, although the high end inverter models can safely power almost anything including sensitive electronic test equipment, there are still some appliances and electrical devices that can be damaged when connected to a lower cost square wave or modified sine wave inverter. If you will use a lower cost non sine wave inverter, you should avoid powering the following electrical appliances from these inverters unless you know for sure they will work safely:

- Laser copiers and laser printers. You may want to substitute with an ink jet or dot matrix printer instead.

- Any battery charger including “trickle” chargers, charger stands for cellular phones and pagers, and charger stands for portable drills and screwdrivers. If these are required, consider plugging these chargers into outlets only powered from the utility grid or a generator.

- AM band radios. These may work but they will have very objectionable noise and hum.

- Most X-10 remote control devices, especially light switches and outdoor motion sensor controlled lights. Also, some X-10 wall switches use an electronic solid state relay that turns all lights on if there is a momentary voltage drop when transferring power from the grid to a generator or inverter. Some X-10 wall switches will actually be hot to the touch when connected to a non sine wave inverter.

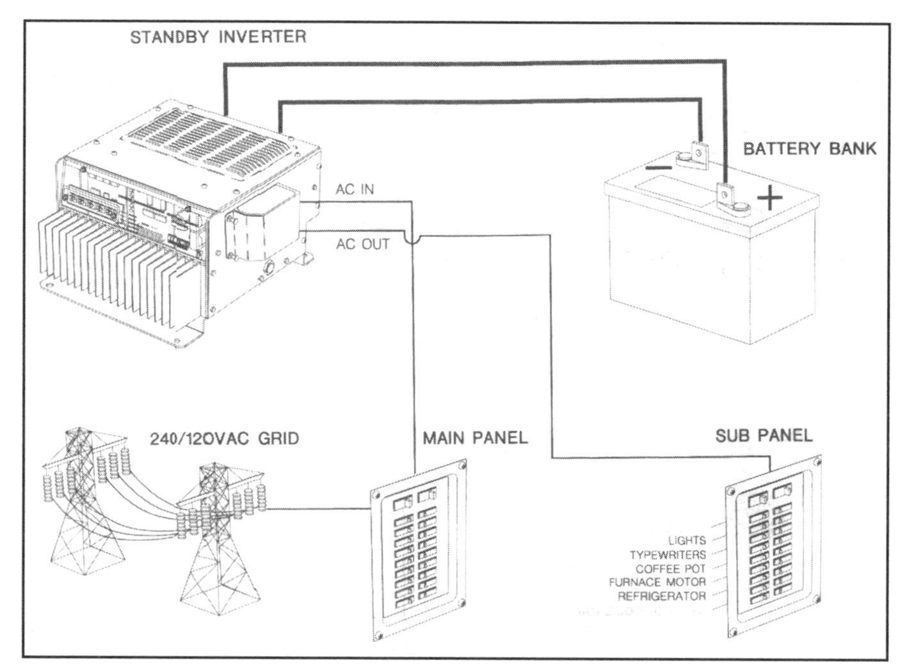

Figure 9 - 2 Basic battery inverter backup power system showing use of sub-panel to feed critical electrical loads. Drawing courtesy of Trace Engineering.

INVERTER SIZES: Inverters are available in many different capacities and battery voltages, but experience has identified several ideal voltage-load combinations for residential systems. The smallest useful inverter size is the 400 to 800 watt 12 volt DC basic model usually serving only one AC load. An inverter this size can power a deskto computer, fax machine, and a fluorescent light fixture usually at the same time. It could also be used to power a 1/4" capacity electric drill, but will be too small to operate a power saw, microwave oven, or vacuum cleaner. This inverter is approximately the size of a small shoe box.

The next inverter size range is 1,500 to 2,500 watts, which is available for 12 volt and 24 volt battery systems. A 2,500 watt inverter has the capacity to operate a ½ HP well pump, clothes washer, microwave oven, vacuum cleaner, or power saw, but only one at a time. By taking care to only operate the lights and one major appliance or tool at the same time, this inverter would be a great choice for an off-grid weekend cabin or emergency power for a home office. This inverter is approximately the size of a small microwave oven.

The next inverter size is in the 4,000 to 5,000 watt range. An inverter this size is not available for 12 volt battery systems due to extremely high currents that would be required at this low voltage. The most common battery voltages for this size inverter are 24 or 48 volt. An inverter with this load capacity will be approximately the size of two small microwave ovens end-by-end. This inverter is extremely heavy and requires a more substantial wall for support.

A 4,000 watt inverter has enough capacity to power a small home and can operate a well pump and washing machine at the same time, while also supply lighting for all rooms. Most residential inverters produce only a 120 volt AC output, not 240 volt. A large home needing to operate many different appliances at the same time may require two identical 4,000 watt inverters to split up the individual loads.

Although most off-grid homes are designed to use only 120 volt AC appliances, if you must operate 240 volt well pumps or shop equipment, some models of 120 volt inverters can be "stacked". These inverters have electronic interconnects to synchronize each unit's output to create 240 volts. It is also possible to use a 120 volt / 240 volt step up transformer if you only have one high voltage load that must be supplied.

INVERTER INPUT CURRENT AND VOLTAGE: In earlier discussions, we learned that regardless of voltage; watts = amps x volts; and the larger the voltage, the lower the current. At 120 volts, a 100 watt light bulb will convert 0.83 amps of current to light and heat energy.

$$\text{WATTS} = \text{AMPS} \times \text{VOLTS}$$

$$\text{amps} = \frac{\text{watts}}{\text{volts}} = \frac{100}{120} = 0.83 \text{ amps}$$

If the 120 volts to power the light was produced from an inverter connected to a 12 volt battery, the current would be:

$$\text{amps} = \frac{\text{watts}}{\text{volts}} = \frac{100}{12} = 8.3 \text{ amps}$$

This is ten times the current flow at 120 volts; however, the energy consumed remains the same if we omit the efficiency losses of the inverter and battery charging process.

$$\text{watts} = \text{amps} \times \text{volts}$$
$$\text{watts} = (8.3)(12) = 100 \text{ watts!}$$

The wire size between the power source and the light remains the same regardless of whether the inverter or utility grid is supplying the power, as long as the voltage remains the same. On the input side of the inverter however, the wire must carry ten times the current when connected to a 12 volt battery system. If the same load is used, but connected to a 24 volt DC input inverter, the current flow drops to 4.6 amps and a smaller wire can be used.

$$\text{amps} = \frac{\text{watts}}{\text{volts}} = \frac{100}{24} = 4.6 \text{ amps}$$

Note the system watts remain the same.

$$\text{watts} = \text{amps x volts}$$
$$\text{watts} = (4.6)(24)$$
$$\text{watts} = 100 \text{ watts}$$

High current flows not only affect wire sizes, they also affect the sizes of circuit breakers, power transistors, resistors, and all electronic components of a modern alternative energy system inverter. For most inverter brands, 2600 watts capacity is the upper limit load that can be produced continuously without unacceptable heat damage on a 12 volt battery system.

$$\frac{2600 \text{ watts}}{12 \text{ volts}} = 208 \text{ amps}$$

In order to provide 12 volts to the inverter and obtain a reasonable period of operating hours, we may need four or more 6 volt golf cart batteries wired in series parallel to produce the 12 volts as shown in Figure 9 - 3.

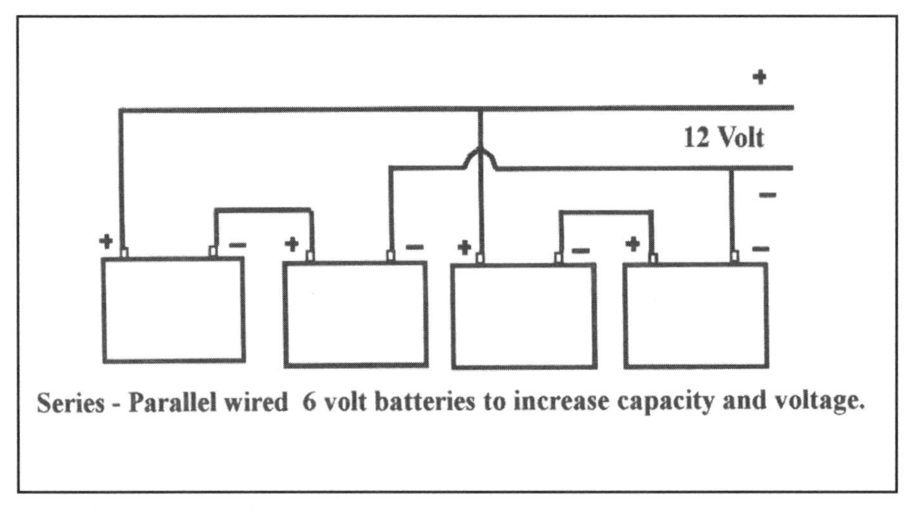

Figure 9 - 3 Using four 6 volt batteries for higher capacity 12 volt output.

If we were to rewire this four battery system, we can provide 24 volts, even though the same total amp hour capacity remains unchanged. Using a 24 volt DC inverter, our current through the battery wires and inverter components will be:

$$\frac{2600 \text{ watts}}{24 \text{ volts}} = 104 \text{ amps}$$

This is half the current flow that we calculated at 12 volts. If we would hold the current flow the same as the 12 volt design the total watts now are:

$$\text{watts} = \text{amps x volts}$$
$$\text{watts} = (208 \text{ amps})(24)$$
$$= 4992 \text{ watts!}$$

This is why an inverter larger than 2600 watts usually uses a 24 volt battery bank. Figure 9-4 shows how 6 volt batteries can be wired in series to provide 24 volts.

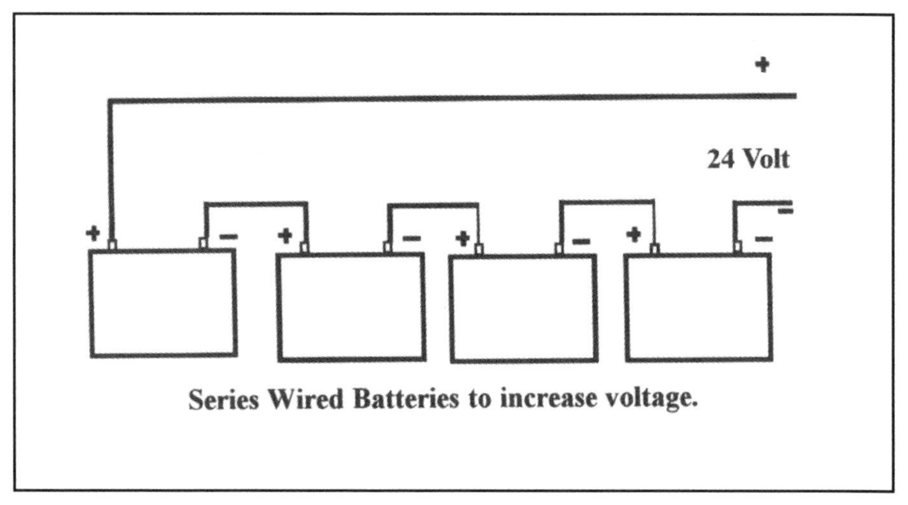

Figure 9 - 4 Using four 6 volt batteries for 24 volt output.

Inverters are also manufactured to operate on 36, 48, and 120 volt DC inputs to further reduce battery input current and wire sizes. For most small cabins, one 12 volt DC inverter rated at 2,000 to 2,600 watts will be sufficient. Most energy efficient off-grid homes will be able to operate from a 24 volt DC inverter rated at 4,000 watts not including air conditioning loads.

Solar or off-grid homes requiring a higher system capacity will usually split the electrical loads between two inverters which also provides 240 volt output for large motor loads when the inverters are designed for interconnection. If your system load requirements are beyond the capacity of two 4,000 watt inverters, you will probably need a custom engineered system that uses an industrial grade inverter operating from a 48, 96 or 120 volt DC battery bank to keep current flow and related wire and components to a reasonable size. A system this large should be designed by very experienced solar design engineers and all systems should be installed by licensed electricians.

PACKAGED INVERTERS: Not all solar or backup power systems require extensive installation wiring. Several self contained inverters are available that include the batteries, inverter, circuit breakers, and charger together in a pre-wired module.

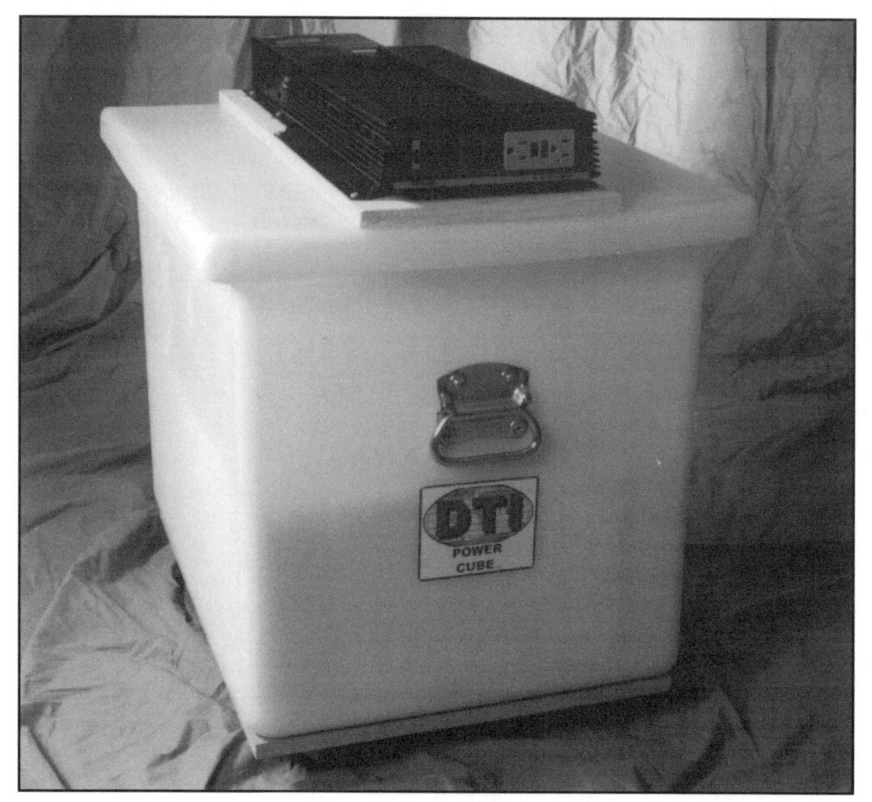

Figure 9 - 5 Packaged inverter and battery module.

Figure 9 - 5 shows a self contained inverter from Dunimis Technology that is available in 800 to 1,500 watt capacities, with several different battery types and charger options, mounted in a leak proof pre-wired enclosure. Since units like this are self contained, they can be charged from any utility outlet, then loaded in a car and taken to a weekend cabin or job site.

There are more and more of these portable self-contained systems now coming into the market, but they all must limit the time they can power your appliances or they would be too heavy to move. Be sure to carefully review published capacity and storage ratings before purchasing a lower priced model having a very small battery bank.

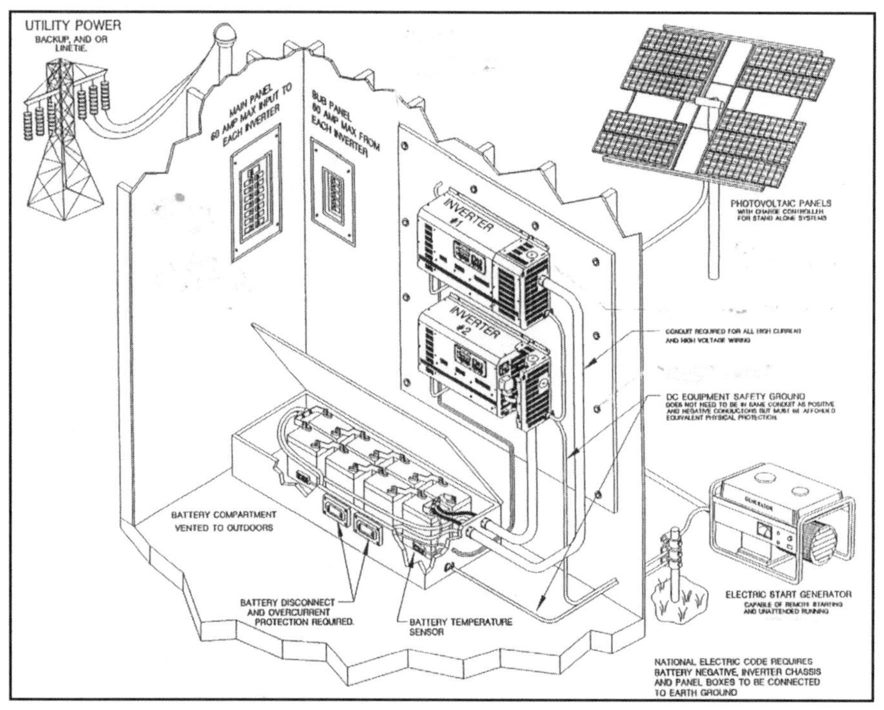

Figure 9 - 6 Large Residential 100% backup power system using two inverters, solar array and generator. Drawing courtesy of Trace Engineering.

LARGER VOLTAGE LOADS: In this text I have referred to several larger appliances and well pumps designed to operate on 220 volt AC utility power. The 220 volt reference is generic, as many utility grid supplied power distribution systems may actually deliver 208 volts, 220 volts, 230 volts, or as much as 240 volts, depending on the incoming power transformer and if single or 3-phase primary power is available. To reduce costs and electrical wire size, most well pumps and shop tools having larger than 1/2 HP motors are only available with 220 volt wiring service connections. If you need to utilize this equipment, you will need two inverters as shown in Figure 9 - 6, or will only be able to operate these higher voltage loads when the generator is running. When using two 120 volt inverters to obtain a 220/240 volt output, the inverters must be designed to allow this interconnection. A communications buss between the two inverters for this intercommunication is also required.

A 120 to 240 volt step up transformer is also available to allow operating a higher voltage appliance from a standard 120 volt generator or inverter.

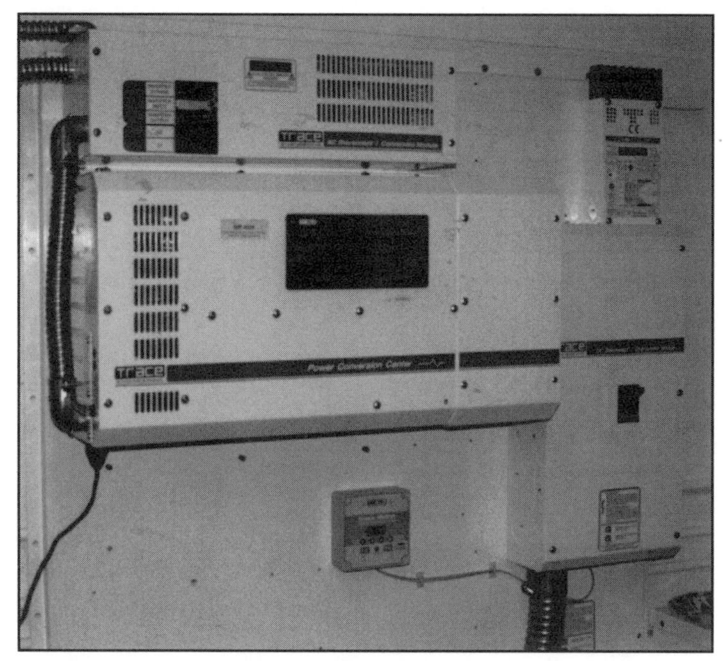

Figure 9 - 7 Trace 120 volt 4 kW inverter with top mounted self-contained 120/240 volt step-up transformer and safety disconnect package.

On smaller systems when it is not justified to add a second inverter just to power a 240 volt load, most inverter manufacturers now offer a step-up transformer that converts the 120 volt output of a single inverter to 240 volt. Figure 9 - 7 above shows a single Trace 4 kW 120 volt output inverter with a top mounted options package that includes this pre-wired transformer and both input and output circuit breakers.

CHAPTER X
SOLAR INSOLATION AND PV ARRAYS

The measurement of solar radiation striking the earth's surface after passing through the atmosphere is called the solar "insolation value" for the location. I hope we all can still remember from junior high science class that the earth revolves around the sun in an orbit requiring 365 1/4 days to complete.

If the axis of the earth was "vertical" in reference to this yearly orbit, the amount of sunlight striking the earth's surface would be the same every day of the year for the same hour of the day. In reality, this is true for only two days of the year; near March 21, the vernal or spring equinox, and near September 21, the autumnal equinox, making day and night of equal length all over the world. For the remaining 363 days, this solar radiation constantly changes because the earth is actually tilted 23 1/2 degrees on its' spinning axis and only during the two equinoxes will this tilt angle align with the circular orbit around the sun which results in an equal day and night.

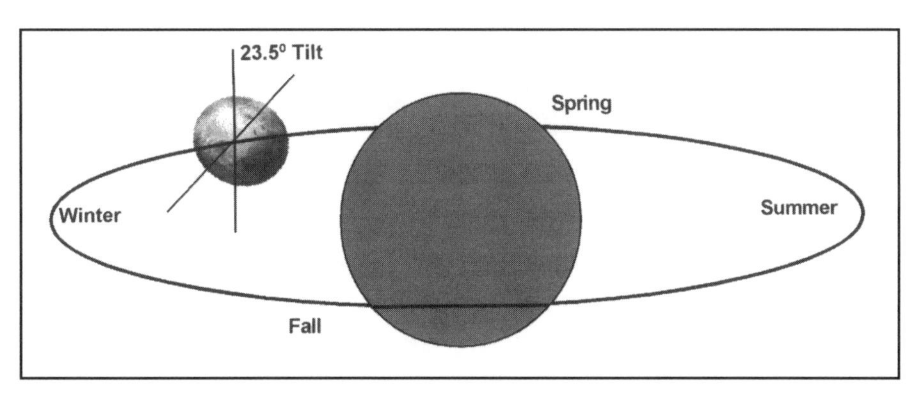

Figure 10 - 1 Changing seasons caused by rotation of Earth around the Sun.

Most solar design books include computer generated monthly solar insolation tables based on Earth latitudes. Keep in mind these average values must be modified to account for your local weather conditions. Knowing this maximum solar energy available for a given latitude however, is a good starting point in determining how many solar modules will be required for a given location.

This insolation data is usually provided only for a horizontal flat unshaded surface, measured in BTU/ft.^{2} for solar heating, and kW/M^2 for solar photovoltaic electric applications. When we add a tilt angle for the flat surface into these calculations and face the surface due south, these total values will change significantly.

As the tilt angle of the sunlit surface becomes steeper, the monthly insolation values for the summer months when the sun is higher in the sky will decrease slightly, while the winter months will increase drastically. These monthly increases and decreases will usually balance out for most site locations when using a tilt angle equal to the site latitude and facing due south.

For solar heating applications, a higher tilt angle is usually necessary for maximum winter performance, and the lower summer efficiency caused by a steeper tilt angle lowers heating output during months with little or no heating is required. In extreme northern latitudes near Canada, mounting the modules on a vertical wall may prove more cost effective, and will result in less problems with snow buildup. In the extreme southern latitudes of Florida and California, most solar pool heating and domestic hot water modules are installed horizontally on flat roofs.

For solar photovoltaic electric panel applications with approximately the same electric loads throughout the year, a tilt angle equal to the site latitude plus 10 degrees will usually provide the best overall system performance since days are shorter in winter months. For homeowners with some mechanical aptitude, a manually tilted array that allows a low tilt angle during the summer, and a high tilt angle during winter will increase the yearly performance approximately 10%.

Our experience indicates there are so many other factors that affect a solar system's performance, that a non-perfect solar module mounting tilt angle and slightly off-south facing will have no measurable effect on an average sized residential system. A slightly lower system performance can be offset by reduced array mounting expenses, if an existing roof can be used instead of buying the ultimate solar support structure.

On small solar hot water and backup solar photovoltaic applications, a module tilted at latitude plus or minus 10 degrees, and facing true south plus or minus 15 degrees will produce acceptable performance. Larger solar arrays will justify the cost of technical assistance for more exact system sizing and array mounting details.

SOLAR MODULES AND SOLAR ARRAYS: Not all alternative energy systems utilize solar photovoltaic power. Many backup power systems use a generator or the utility grid to charge the battery bank and do not use or need solar modules; however, I am including the following brief description of how a photovoltaic array works for those of you planning to include solar in your home power system.

Whenever we speak of a solar panel or solar module in this text, we are referring to a single individual photovoltaic device, usually having a glass face and aluminum edge frame, and consisting of individual solar cells internally wired together to produce the voltage and current specified by the manufacturer. The individual cells are laminated between the back of the glass front, and a highly moisture resistant vinyl backing material. The module includes some form of weather proof junction box or connectors to allow external interconnection of multiple modules together to form a "solar array."

Some modules are also available with a non-glass front, or the entire assembly may be bonded to a solid aluminum backing without the conventional frame around the module like those shown in Figure 10 - 2. In most cases, these non-glass modules will have a shorter life expectancy and manufacturer warranty, but they offer the advantage of flexibility and added ruggedness for mounting on irregular surfaces like the decks of boats and roofs of RV vehicles. Solar modules come in all shapes and sizes; however, most residential photovoltaic modules have widths between 12" and 26", and lengths between 39" and 48".

Figure 10 - 2 Examples of small, semi-flexible recreational solar photovoltaic modules.

HIGHER VOLTAGE MODULES: Sometimes the manufacturer will make the same exact module with two or more additional individual cells to increase the voltage output. For example, a standard module will usually have an output voltage under load of around 15 volts for charging a 12 volt battery system. This slightly higher voltage provides the required "push" to move the current charge into the battery which is at a lower voltage, and also overcome any additional electrical resistance in the circuit.

Although elevated ambient temperatures will not normally damage a solar module, they will reduce the output voltage. Many system designers prefer to use a higher output voltage version of a manufacturers module for very hot locations like the extreme southern latitudes, desert areas, and above large flat industrial heat reflecting roofs subject to long summer afternoons of intense direct sun.

A higher output voltage module in a high ambient temperature environment, will produce the same current flow to the batteries that a standard module not operating in such a hostile environment. Using this higher voltage module in a less high temperature application does not increase the amount of energy collected, since the voltage is already higher than the required battery charging voltage and the total current available did not change. Since the higher voltage modules are slightly larger and contain extra cells, these modules require more of the same material to manufacture and have a higher cost.

The cost benefit of the higher temperature modules may not be justified on a typical remote home in the higher mountains or cooler wooded valley wilderness sites.

PHOTOVOLTAIC CELL TYPES: The original solar module construction required "growing" a long crystal of silicon, approximately three to four inches in diameter, which was sliced into very thin silicon wafers. These silicon wafers would then have electrical conductors called "fingers," hand soldered or plated to the silicon surface. Since any metal conductors attached to the surface of the silicon wafer would shade the silicon under the conductor from sunlight, these electrical conductors needed to be extremely narrow and thin. This was obviously very time consuming to manufacture, and the slicing process usually wasted more of the very expensive silicon in the width of the cut, than was used in the actual resulting cell.

Some solar modules are still manufactured using this technique because this industrial grade silicon cell construction still has the highest efficiency, and produces almost the same output regardless of any off axis sun angles encountered throughout the day. Many industrial and outer space applications still demand this higher priced individual single crystal silicon wafer cell construction.

To reduce solar module costs, many manufacturers now use a "polycrystalline" silicon material. With this process, the silicon is cast into cubes that have a very large silicon grain structure. This makes each crystal grain boundary act like an individual cell. The surface coatings applied to provide selective absorption of the sunlight for this lower cost cell construction results in a solar module with an overall blue appearance. You will probably notice this type of cell in your solar powered calculators and radios. This process is much less expensive than "growing" a large silicon crystal, and still results in very good cell performance and life expectancy. The off axis performance however, when the sun is not directly vertical to the module surface is slightly less than modules using the individual silicon wafer cell construction.

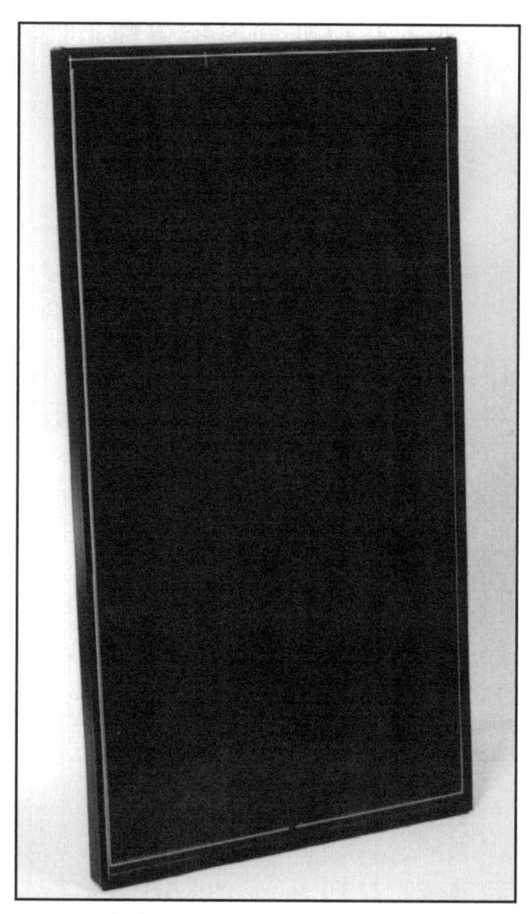

Figure 10 - 3 Large thin film photovoltaic module design.

We spoke earlier about how a solar module is made up of individual solar cells but for some solar modules, each individual cell is almost indistinguishable from their neighbor since they are etched out of the silicon material which has been deposited on the glass surface chemically.

This manufacturing process called "amorphous" or "thin film" cell technology allows depositing the silicon material in a very thin uniform layer on the glass cover plate by exposing it to various rare gasses inside a pressurized chamber. This results in a continuous thin film of silicon which is later divided into individual solar cells using a laser to cut the silicon film. Plated onto the glass cover plate the resulting glass solar module appears to be a uniform dark gray or black painted glass, and the tiny laser drawn circuits and individual solar cells can only be seen at very close distance.

Figure 10 - 3 shows one of these larger solar modules made by a fully automated gas deposition silicon film process. This solar module construction is becoming popular for larger commercial applications since the modules can be produced in sizes approximately the size of a typical residential door. These larger modules usually have output voltages above 90 volts DC, and the larger individual module size results in fewer inter-panel junction boxes, less above roof wiring exposed to weather, and requires less individual mounting hardware.

Solar modules using thin film technology usually cost less per module than polycrystalline modules, but more modules and roof area will be required due to their lower operating efficiency. The cost per watt generated, which is used to compare the cost effectiveness of modules, will usually still be in the same range as modules made with conventional cell construction.

A more detailed listing of the most popular solar photovoltaic panels and their physical dimensions is included in TABLE #9 in the Appendix.

SOLAR MODULES AS ROOFS: A recent trend in solar arrays is to reduce system costs by using the solar module as part of the roofing material itself. By eliminating roofing materials, the roofing material savings can offset part of the still relatively high cost of solar modules. Some module manufacturers are providing modified frames around the solar module that allow direct attachment to the roof rafters, with the remaining gaps between each module calking and sealed to keep out rain and snow.

Glass solar modules can be manufactured with a semi-transparent backing material which allows the solar module to be a form of skylight, providing passive space heating, day lighting, and electrical power generation at the same time.

The newest idea in replacing roofing materials is the solar shingle as shown in Figure 10 - 4. This is not however, a flexible asphalt shingle that happens to generate electricity.

Figure 10 - 4 Solar roofing shingle and interconnect plug. Courtesy Atlantis Energy Systems.

The solar shingle design uses a hard cement fiberboard with the upper half blank to fit under the row of shingles above.

This upper area also includes holes to nail into the plywood roof sheeting for attachment, and a very low profile electrical socket. An interconnect plug and wire that will just reach the next shingle provides the only electrical connection needed. The plug and socket are designed to connect by hand, and form a water tight junction due to the special plug design.

The lower half of the solar shingle has solar cells bonded to a metal plate laminated to the shingle surface. At the end of each row of solar shingles is a low profile wiring conduit to electrically tie each row to the next. This wiring procedure allows the roof array to continue to operate even if several individual solar shingles are damaged or electrically shorted. Although still somewhat expensive to produce, the solar shingles can be installed by a conventional roofing crew without any wiring experience or electrician's license which allows some additional installation cost savings.

After installation, the overlap of each row of shingles covers over all wiring and cell interconnects, and at a distance, appears to be a conventional asphalt shingled roof.

Changing architectural design needs and clients wanting solar arrays that do not stand out in their neighborhoods will probably drive the market in the direction of roof integrated solar arrays for solar installations in urban areas. Japan is a major user of this roof integrated technology. This will also make it much easier to retrofit a large solar array onto existing homes that already have an existing large south facing roof at the correct angle, that is not shaded by nearby trees.

HOW DOES IT WORK? Those of you planning to use solar modules as your source of your electrical energy do not need to know how they work to benefit from their magic, but you may want to at least understand the basics before talking to curious neighbors.

To fully understand the photovoltaic process requires the knowledge of solid state physics, atomic energy levels, and quantum mechanics. But the basic concept is - a solar cell is nothing more than an electronic diode or check valve which allows an electric current to pass "backward" through it when it is not exposed to sunlight. The solar module will "drain" the charge from the battery that it charged during the day if it is connected directly to the battery without a charge controller or diode to block this reverse current flow at night.

When sunlight shines onto the solar cell, the photons in the sunlight energize the electrons in the atoms of the silicon material, which overcomes the natural electrical diode effect and causes an electrical flow to take place in the "forward" direction. Almost any silicon cell electrical junction will generate slightly over 1/2 volt when exposed to direct sun, regardless of the size or shape of the individual cell. The increased cell size however, will increase the current capacity.

One of the most unusual electrical characteristics of a solar module is that there is little or no damage if exposed to sunlight and the terminals are not connected to a load. In addition, there will be little or no damage if exposed to sunlight and the terminals are "shorted" together. Since the photons in the sunlight only cause a "flow" of electrons to take place inside the silicon cell structure, no material is actually consumed. Theoretically, a solar module should have an infinite life; however, module failures do occur, since long term sun exposure and constant thermal expansion and contraction causes the backing sealant to pull away from the glass. This allows moisture to enter between the glass and cell backing and corrode the cell interconnect wiring. A solar module made from quality materials should provide reliable service for over 25-years, and many now carry a 20-year manufacturers warranty.

CHAPTER XI
SOLAR BATTERY CHARGING

Figure 11 - 1 Residential solar photovoltaic array using multiple Arco 12" x 48" modules.

Any battery based system can include a photovoltaic solar array to keep the battery charged. I am always asked how many solar modules are needed to charge a battery and theanswer is linked to many variables. Depending on the local weather, cloud cover, ambient temperature, site altitude, season, location latitude, solar module azimuth orientation and tilt angle, the same solar module will generate a different current output for each minute of the day, although the voltage will remain fairly constant or rise only slightly. It should be obvious that this analysis requires some fairly detailed calculations to arrive at the most optimum combination of battery size, solar array size, and array orientation for your site.

On larger commercial and institutional solar projects, sophisticated computer modeling programs compare the proposed system design output for each hour of the year using different combinations of mounting positions and battery size. Thanks to the many residential systems already installed with similar lighting and appliance loads, we can predict smaller system performance with reasonable accuracy as long as several basic installation guidelines are followed.

SOLAR CHARGING VS. GENERATOR CHARGING: Any time you can add a solar photovoltaic array to an existing emergency generator or battery based residential power system, the run time of the generator will be reduced. This reduces the need for refueling, generator maintenance, fumes, and noise. The larger the array, the more days between generator operation up to a point where the generator is non-existent. Most off-grid homes still have a backup propane or diesel fueled generator for those occasional large motor loads including a weekly wash day when the well pump, washer, dryer, and clothes iron may be operating at the same time.

Increasing the solar array and battery bank size to build up enough stored power for these once a week loads, only to have the batteries discharged in a matter of several hours, is much less cost effective than operating the generator and feeding these loads directly without going through the inverter and batteries. In addition, there can be up to 20% energy loss in the charging and inverter process that you will not have when feeding these large motor loads directly.

ARRAY WIRING: Each solar module has a nominal output voltage that determines the battery voltage that can be connected. Like batteries, solar photovoltaic modules can be wired in series to increase circuit voltage, wired in parallel to increase circuit amperage capacity, or wired in series-parallel to increase both. Solar arrays with eight or more modules will usually include a "combiner box" located near the array which provides a method to connect the many smaller wires from each module to the larger feed wires leading down to the charge controller. Commercially made combiner boxes for larger numbers of modules usually also provide several manual switches to allow dis-connecting individual groups of modules to make testing and repairs on part of the array without shutting down the entire system. Some combiner boxes may also include a lighting arrester device and grounding terminal, which may be desirable for many locations where lighting strikes are common.

CHARGE CONTROLLER:

Regardless of the solar photovoltaic array size and capacity, the electrical energy being generated from the array cannot be fed directly into the battery bank. Several safety and system performance issues dictate the need for a solar charge controller as shown in Figure 11-2.

Figure 11 - 2 Solar charge controller to regulate the solar charging process of a battery bank.

Although the voltage is fairly constant for a solar array regardless of changing solar angles and cloud cover, the current flow can change dramatically. As the battery bank reaches full charge, the current flow from the solar array must be reduced or the batteries will begin to out-gas and lose water. At night, the solar array becomes a system load and current will try to flow out of the battery and back through the solar array. The charge controller provides all of these safety control features and some models also include a digital display of array voltage, battery voltage, and charge rate. Some controller manufacturers believe "pulsing" the charge current into the batteries instead of a constant charge rate can increase battery life, and their charge controllers provide this feature.

ARRAY GROUND FAULT: If your solar photovoltaic array will be mounted on the roof of an occupied dwelling, the *National Electric Code* requires the wiring from the array entering the house to have a DC rated ground fault device. This device is designed to sense when there is a 1 amp flow in the ground wire and disconnect the positive and negative leads from the array to stop current flow that could cause a fire or shock. There was some conflict between the actual *1996 Electric Code* and the *National Electric Code Handbook* regarding this ground fault device and if it needed to "open" the positive and negative leads from the solar array or shunt them together in the event of a fault condition. This has been clarified with the 1999 Code requiring the ground fault device to open on fault condition. The code does not require this safety device for solar arrays on poles or ground mounted.

BYPASS DIODE: A diode in an electric circuit provides the same function as a check value in a plumbing circuit - to allow flow in only one direction. When more than one solar photovoltaic module is wired in series to increase array voltage, any shading or defect of one module reduces or blocks the flow of current from the other modules in that series "string." Like the Christmas tree light string that will not operate due to one burned out bulb, a problem with one module effects all other modules. Most solar modules manufacturers now provide a bypass diode wired in parallel across the output terminals of each solar module. Under normal conditions the current flow is through the protected module and not the diode, but any drop in voltage output of the protected module causes the current from the balance of the array string to "bypass" the shaded or defective module.

BLOCKING DIODE: Unlike the bypass diode which is intended to allow a string of modules to continue operation with a shaded or defective module, the blocking diode is used to prevent permanent damage of modules caused by a reverse current flow. When more than one string of individual solar modules are wired in parallel with other strings of solar modules, there is an increased risk of current flow from one string back feeding into another string of modules. This condition usually occurs when several modules in an array are shaded by chimneys, clouds, or trees. When this occurs the shaded modules will try to absorb current from the parallel wired modules. On smaller low voltage arrays this is usually not a problem, but on larger higher voltage series-parallel arrays the backward flow of current can heat up and permanently damage the shaded or under producing solar module.

Although newer thin film technology modules are less sensitive to reverse current flow than crystalline type modules, most designers now include a blocking diode on the positive lead of each string that is wired in parallel with other strings. Since the diode in this application is wired in series with the string of individual solar modules, its voltage drop will consume some of the energy generated by the solar modules. A string blocking diode must be carefully selected to handle the maximum currents and voltages expected with minimum energy loss, and they should be mounted on a heat sink to reduce any heat buildup.

ARRAYS FOR OUTBUILDING LIGHTING: A solar photovoltaic system does not need to be large and complex. There are many outbuildings on a typical farm or ranch that are too far from the main residence to justify long electrical power lines. Many of these outbuildings only require interior or exterior lighting for occasional use, and there are no other electrical loads.

Figure 11 - 3 Solar powered outdoor light kit.

Installing a solar module, charge controller, and gel cell battery that will not require any maintenance, is an ideal way to power a fluorescent interior or exterior light for a remote building.

Figure 11 - 3 shows how easy it is to provide a photocell controlled area light outside a small remote outbuilding. All that is required is a solar module, gel cell battery, moisture proof 7 to 15 watt fluorescent light, and a solar charge controller that also uses the solar module as a "photocell" at night to signal when it is dark. Some charge controllers also include a timer that will turn off the lights after a set time period has past.

The gel cell battery eliminates the problem of battery maintenance, but should be in an insulated and vented enclosure if it will be subject to freezing temperatures. Since the light will have longer night hours of operation than the solar array will have daylight, be sure the solar array has 4 to 5 times the wattage rating of the light, and the battery can store several days energy usage. This weekend solar powered lighting project is an excellent place to start your "experiment" with solar systems, which still has a very useful purpose.

ARRAYS FOR BATTERY CHARGING: If you are planning a solar battery charging system that will only provide lighting for a small weekend cabin, you will probable want to useonly 12 volt DC lighting and not require a DC to AC inverter. A system this small will probable use two to four golf cart type batteries and require one solar module in the 50 watt range for each battery in the system. To save costs, many of these smaller arrays are attached directly to the existing roof using brackets that keep the modules at the same angle as the roof. These smaller systems may also face off south if the existing roof is not orientated directly south. As the system size and cost increases, array performance increases from proper tilt angle and orientation will justify the cost of array racks or roof modifications.

Systems with four to twelve solar modules may include a small inverter to power AC loads including a television, stereo, micro-wave oven, or small power tools. Systems this size may have slightly more than one golf cart size battery per solar module, which will provide more days of cloudy weather backup operation.

Solar arrays with twelve to sixteen solar modules in the 50 watt per module range can usually charge two golf cart size batteries per module for most locations in the continental United States. Arrays and battery banks this size may be wired for 24 volt operation, and include a 4000 watt inverter to power the lighting, small appliances and entertainment equipment, and a deep well pump. Systems in this size range are near the break point of switching to a larger industrial rated battery and could include more than one charge controller and inverter depending on system loads and array current capacity.

Although we have been discussing charging batteries located inside a residence with a solar array, it should be noted that more and more solar photovoltaic systems are being installed on homes that remain connected to the electric grid. For these applications, there will still be an inverter to convert the DC voltage of the solar array into AC voltage to match the utility grid, but the utility grid becomes the "battery" or load for this "green power." The utility meter for these installations are designed to meter forward and backward current flows, and any electrical power generated by the solar array that is not immediately consumed in the home is sold back to the utility grid. The major problem with this system is that without a battery, the system will not function if the grid goes down since there is no buffer to balance the current flows when the load does not match the array output.

These "grid tied" systems also have very strict guidelines for safety disconnects and controls to prevent a continued back feed into a power line that was thought to be de-energized by a maintenance crew for repair.

SIZING A PV ARRAY: I stated at the beginning of this text that it was not my intent to provide solar system sizing analysis as there are several excellent texts and standardized design forms available including maps showing solar insolation for each state and country. If you are planning to install a large photovoltaic array, please refer to these sources to supplement this text. To assist with the comparison of solar modules from different manufacturers, Table #9 in the Appendix lists the most common higher quality off-grid solar modules including their physical dimensions and electrical characteristics.

In real life however, most people select a solar system based on their budget and standard package sizes, and not on what lengthy calculations and computer projections recommend. For example, solar "kits" are usually sold in two module, four module, eight module, twelve module, and sixteen module configurations. Most of these kits will use individual modules having a 40 to 60 watt maximum output rating per module depending on physical size and manufacturer.

In other words, instead of selecting an array size based on your actual electrical needs, it is assumed the array can never supply all of your loads and the utility grid or generator will be there to take up the slack. Additionally, for off-grid applications this may require delaying the use of some appliances until sunny days.

Using this "backward engineering" approach, our goal is to estimate the average useful energy we can expect for a given array size. This simplified methodology is given in TABLE 10 of the Appendix.

ARRAY MOUNTING: Most people think of a solar array as adding weight to an existing roof and this is their primary structural concern. Adding weight to any roof truss system does increase truss loading, but most solar arrays have a very low weight per square foot surface area that is well within the safety limits of almost all well built roof and truss systems. The real structural impact that must be addressed for a planned solar array is wind uplift. This is the result of wind pressure trying to pull the solar modules up and off the roof.

Since solar modules are designed to withstand these forces, our primary concern is the module assembly trying to pull the mounting bolts, lag screws, or support mounts out of the roof. These uplift forces are even greater within three feet of the roof edge or roof eve. Follow all of the module manufacturer's mounting instructions and be sure module supports are near or directly over rafters. Do not rely on randomly spaced screws in 1/2" plywood sheeting to withstand these uplift forces, and never install a solar array on a roof having aging shingles that will soon need to be replaced.

CHAPTER XII
LIFESTYLE CHANGES

I am often asked about living in an off-grid home; especially by a concerned spouse. There are some real differences between living on and off the grid; some good, some not so good, but this lifestyle can be very fulfilling. For a person living their entire life in an apartment in an urban area, moving to the country can be very unsettling. For example, waking up one morning and discovering the absence of water coming out of the shower faucet may result in running for the phone to call the building manager before you realize its up to you to solve the problem. On the other hand, you will most likely not get mugged while sitting on your front porch viewing a beautiful sunset or a majestic sky full of glittering stars. Like the owner of the rural log home shown in Figure 12 - 1 who converted her home to photovoltaic power, you too will no longer worry about power interruptions.

Figure 12 - 1 Rural packaged log home using ground mounted solar array.

Providing your own power gives you the freedom to live where others cannot, and many large tracts of land are still available at low cost due to real estate agents unable to locate families willing to build in areas not served by power lines. The off-grid lifestyle has the flavor of what I believe the early American pioneer must have felt. There is a feeling that your shelter, your food, your heat, and your comfort are from your own efforts, with much less dependence on others for your day to day needs. There are also many levels of this dependence/independence relationship with the land.

The "back to nature" movement has always drawn individuals who wanted to be isolated from everything we call society. This segment of the off-grid population are at the extreme end of self- sufficiency, sometimes choosing to grow all their own food, canning foods for the winter, raising chickens for eggs, and cows, sheep, or pigs for meat, pumping their water from a hand dug well, using kerosene lanterns or candles for light, and doing without indoor plumbing and all electric appliances. This is not to praise or condemn these individuals, but to acknowledge that there will always be some who feel this minimalist lifestyle is right for them. Chances are, if you have read to this point, you would find this a somewhat primitive level of hearth and home, and this extreme is not what you have in mind when considering an off-grid home.

I call the larger middle group of alternative power adventurers - the "reliability conscience." This group will most likely want to remain connected to the power lines, but do not like the feeling of total dependence on the utility company. They may live in an area where one or two day power outages are common during winter storms or spring rains, and they just want some level of emergency backup power to see them through until the lights come back on.

Many people in this group also have a strong feeling of energy conservation and want to minimize their energy usage. They may have already considered installing a solar hot water heater and have replaced most of their incandescent bulbs with compact fluorescent lamps. Some people in this group want to reduce their electric and utility usage to a minimum but still expect to have a basic monthly electric bill. This group can easily afford their monthly utility bills, but just do not like wasting energy or feeling dependent on others for their energy needs.

Perhaps you are in this middle group and are considering a backup generator or small photovoltaic array "just to experiment". Maybe your budget will not allow the investment of 15% to 20% of the value of your home in an alternative energy system, but you want to at least be able to keep some lighting and appliances operating during an emergency. This is not only a worthwhile goal, it is also the starting point for families who later want to move up to building an off-grid weekend retreat or vacation home. Think of this level as a time to get your feet wet, learn the ropes, and show other family members what this is all about.

I call the final group of off-grid bound adventurers the "contemporary independent." This group has the desire to live in the more isolated country or mountains, but still want a fairly normal lifestyle that includes all of the time saving conveniences we now consider necessities. They want a small garden, but still plan to purchase most food items in a nearby town which requires refrigeration equipment. They are willing to give up going to the theater or stadium each weekend, but want to keep their satellite dish, VCR and color television.

They are willing to use ceiling fans, wood stoves, and open windows for temperature and comfort control, but still want heat if they do not feel like chopping wood or building a fire. They are willing to monitor their system and always plan ahead before operating larger appliances, but they still want these appliances.

Energy awareness is one of the most important aspects of alternative power living. Eliminating the power company makes you the power company, and also makes you the power plant manager and maintenance director.

Generators require periodic oil changes, servicing, and a constant source of clean fuel. Solar modules are fairly maintenance free, but will require some exercise with a broom after a heavy snow. Batteries need to be kept clean and topped off with distilled water. Lead acid batteries also require an occasional equalize charge every few months and all batteries must be replaced every few years. The maintenance aspect of an alternative power lifestyle may be beginning to sound time consuming, but in reality most of these items can be preformed in a few minutes. The hard part of this maintenance program is keeping track of when these tasks should be performed. Even though you may still be connected to the utility grid, to maximize the efficiency of any solar array or battery bank requires an efficient use of lighting and appliances.

It is important to maintain a more constant level of electrical usage and not have extreme swings in system demand. Batteries are more efficient when discharged slowly over many hours, and generators do not like rapid fluctuations between idle and full power. Solar modules do not harvest sunlight at night or during storms and wind generators may go days without enough wind to operate. These energy ebbs and flows in your power system may not match your personal energy usage patterns. If you wish to live totally off-grid successfully, you and your family must learn to work with and not against these natural cycles and equipment limitations. If today is laundry day, it may not be the best day to also operate power tools. The goal here is not to eliminate these labor saving conveniences in your life, the goal is to use power efficiently and not run out.

Before embarking on an alternative energy system you should consider your age and health. Let's face reality, most older people would not appreciate cutting firewood for heat, starting a temperamental generator when batteries are low or climbing on their roof to repair a loose wire or adjusting the tilt of a solar module. This does not mean the older population cannot have the freedom of off-grid living, but it does mean their alternative power systems must be as simple and reliable as possible. If you cannot program the date on your VCR or replace a faucet washer, you may not find this lifestyle very forgiving. We recently installed a very large solar power and backup generator system for an 83-year old women who lives alone in a rural area. She is getting along quite well, but she is in good health and has taken the time to learn everything there is to know about her power system components. She enjoys not worrying about the many power outages her younger neighbors experience and knows to check the battery charge meter before doing her laundry.

The final lifestyle consideration for those planning to install an alternative energy system is realizing you will be very unique in your area. It will be many years before these systems are commonplace and this means you will be very popular in your neighborhood if the local utility grid goes down for several days. You may also receive many requests for house tours from local schools and civic organizations. During installation don't be surprised if you experience some resistance with zoning or building code officials who may not be familiar with these systems and are not appreciative of your sense of conservation. However, please remember it is really worth the effort and added piece of mind.

CHAPTER XIII
WHERE DO I START ?

At this point in your reading, many previous chapters may still not fit together in your mind, and you are not sure where to begin on the road to energy independence. I have simplified the process by dividing everything needed into a four step action plan:

STEP ONE: YOUR FIRST ASSIGNMENT - LIGHTING

The first place to start in weaning yourself from the electric grid is to resist the temptation of running down to the local building supply store and buying a generator. Your first assignment is to replace every incandescent lamp in your home with low energy brands including compact fluorescent bulbs or fluorescent tubes.

Halogen PAR 30 and 38 lamps in the 40 to 60 watt range can be substituted for higher wattage incandescent bulbs in recessed ceiling fixtures or those having dimmer controls, since most fluorescent fixtures will be damaged if wired to a dimmer switch. Small halogen reading lamps and task lamps can be added near chairs or tables to reduce the use of multi-lamp ceiling light fixtures, but avoid those having plug in wall transformers.

You will probably find that some fixtures originally installed in your home are inexpensive contractor grade, and are not designed to use the larger compact fluorescent lamps. These fixtures will need to be replaced with fixtures that will. Larger "harps" are inexpensive and available to fit existing table lamps to allow using these larger fluorescent bulbs without changing the shades. Incandescent kitchen ceiling lights should be replaced with one or more two tube, 4 foot T-8 surface mounted fluorescent fixtures.

When replacing new lighting fixtures, shop for those with clear or diamond pattern lens covers, as a frosted fixture lens greatly reduces the light output.

The standard 4 ft. T-12 fluorescent tube is being phased out of production and will be replaced by the smaller diameter T-8 tube in the future. This new T-8 tube is much more efficient and produces more light than the T-12, but requires a different ballast. Although screw-in compact fluorescent lamps can easily replace an incandescent bulb, fixtures designed to use the plug-in 4-pin compact fluorescent lamps are more reliable and efficient, and have longer life than the 2 pin lamps when powered by an inverter based electrical system.

When selecting any fluorescent lamp or tube, always request "warm" color temperature lamps. The older bluish color fluorescent office lights do not belong in a home. You may need to shop for these in a lighting fixture supply house, but the better light quality is worth the slightly higher price. Plan on completing the re-lamping over a period of several weeks with the goal of throwing out all incandescent bulbs. Don't forget to replace the high wattage outdoor flood lights, which now have outdoor style compact fluorescent and halogen substitutes.

You should notice some immediate reduction in your utility bills, assuming the weather has not changed enough to cause an increase demand on the air conditioning or heating system. Halogen lights are excellent in bathrooms and although more expensive than standard incandescent, they will have much longer lamp life and require less wattage than a standard incandescent bulb. When we first moved into our new 3,400 square foot solar home, every light was new and was either fluorescent or halogen. We are now entering our fourth year in this house and we have yet to replace a single light bulb! Although at some point they will reach their design life, it still is nice to not buy a pack of incandescent bulbs every time you go to the grocery store.

I also recommend replacing wall switches in bathrooms, kitchens, and hallways, with infrared motion sensors to turn off lighting in unoccupied rooms. These are an easy, one-for-one replacement, but be sure to turn off the circuit breaker and follow all installation instructions. Some lower cost motion sensors will turn the light on after it was off due to a brief power interruption. This can be very distracting and wastes energy. Some lower cost X-10 remote control light switches also have this problem.

STEP TWO: YOUR SECOND ASSIGNMENT - APPLIANCES

Your next step is to plan on replacing several major appliances during the next six months. Although some large appliances like clothes washers and dryers are very inefficient, they are not used every day and do not need to operate during a power outage. The appliances that should be replaced are those that operate daily, are high energy users, and are over five-years old.

Our most likely suspects are the refrigerator, the hot water heater (if electric), and the dishwasher. These appliances have undergone extensive design changes by manufacturers to make them more efficient to meet new federal energy regulations. They are also quieter, better insulated, and most have a low energy switch mode control. If replacing a perfectly functioning older refrigerator is not acceptable to your spouse or your sense of thrift, how about purchasing a super efficient top load freezer. Several models are available that will keep foods cold for days and will easily operate on a small portable generator.

Some freezers are available with dual voltage compressors and will operate from the AC wall outlet or a 12 volt battery during emergencies. Again, you should see a noticeable drop in your monthly electric bills after these changes are made. If you operate a home office, consider replacing your laser printer with an ink jet model which uses much less electricity, and replace all computers over three-years old with the new "energy star" computers which have a low energy standby mode when not in use. Off-grid applications should select a clothes washer that has both low water and low electricity usage. A clothes line still makes a great clothes dryer.

STEP THREE: REDUCE STANDBY LOADS: From now on avoid buying any electronic device that requires a plug-in transformer. These allow the manufacturer to reduce appliance size and cost, but it places the on/off switch after the power transformer. These devices consume almost the same electricity turned off that they do on, and they are plugged in 24-hours per day. This is also true of appliances having remote controls since some or all of the electronics must remain on to receive the remote turn-on command. Most of the models we tested wasted two to five watts each, which is converted into heat. You will be surprised to find as many as ten of these devices plugged in around your home, and that can represent the same losses as leaving a 50 watt light on 24-hours per day, 365 days a year!

Although you are probably not going to throw all of this equipment away, most will normally last only two to three years, and future replacement purchases should be with equipment having built-in power supplies and a simple wall plug for power. Avoid models with remote control features unless product literature indicates little or no energy is used while "off".

STEP FOUR: BUY YOUR BACKUP POWER EQUIPMENT!

Okay, now it's time to buy your backup power equipment. You have eliminated all unnecessary electric loads and reduced your home's energy requirements to a minimum for all remaining lights and appliances. This allows buying a much smaller system than you would have needed if this was not done first. Reducing the loads first allows generators to be downsized for fuel economy, solar arrays to be smaller, and battery banks to provide longer periods of backup power between charging.

I strongly recommend shopping for a generator at a contractor supply house or generator distributor. Those 4,000 watt portable generators you see at many home builder supply stores are not designed for long term standby service. They are usually very hard to start in cold weather, have small top mounted gas tanks that are dangerous to refill when hot, and have small mufflers causing loud operating noise.

The more expensive 4,000 to 6,000 watt contractor or emergency standby service grade generators have very quiet mufflers and large separately mounted gas tanks. Their longer life cast iron engine blocks have ball bearings instead of cheaper sleeve bearings, and offer needed options like electric starters and duel fuel gas and propane carburetors. Although most portable generators produce both 120 volt and 240 volt output, select a model which provides full generator capacity output at 120 volt. If you anticipate only needing a generator for emergency backup during power outages only one or two days a year, a 3600 RPM contractor grade generator will meet all your requirements and still be reasonably priced. If your area has power outages up to a week long, purchase the lower speed 1800 RPM model. The lower engine speed requires a more expensive and heavier constructed engine, but the lower speed greatly extends generator life, has higher fuel efficiency, and operates much quieter.

Due to less oxygen for complete combustion at higher elevations, all generaters should be derated if located above sea level. Deduct 3% from the total kW output capacity, for each 1,000 feet increase up to 6,000 feet, then 6% per 1,000 feet after 6,000 feet.

> Example: A 6.5 kW gasoline fueled generater will be located at a 7,500 foot elevation. What is its expected output?

$$\frac{(3\%)}{1{,}000}(6{,}000) + \frac{6\%}{1{,}000}(1{,}500) = 0.18 + 0.09 = 27\% \text{ derate}$$

$$(6.5 \text{ kW})(1.00 - 0.27) = 4.75 \text{ kW output}$$

Although still portable, most generators in this size and price range should be permanently mounted as shown in Figure 13 - 1. To increase safety, most people pour a small concrete pad behind the garage or in the back yard to serve as a mounting base for their generator. A "dog house" style generator enclosure can be built to cover the generator and rest on the concrete pad using siding and roofing materials to match your house, but remember to make it removable to allow generator servicing.

Figure 13 - 1 Backup propane fueled generator.

An enclosure will reduce noise, but will need provision for the muffler exhaust and fresh combustion air intake as shown in Figure 13 - 2.

Note the duplex outlet and timer switch to provide a periodic trickle charge of the starting battery. When pouring the concrete pad, be sure to provide a 1" or 1 1/2" electrical conduit up through the concrete near the generator's control panel, as shown in Figure 13 - 2, to route all power and control wiring underground to the house or garage.

Figure 13 - 2 Same generator with "dog house" style removable enclosure and sound proofing.

Since these generators are still portable, we also recommend embedding a large 1/2" steel "eye bolt" in the wet concrete near one corner to attach a padlock and heavy chain visible in Figure 13 - 1, as these generators have been known to walk away from their home.

Although you can run an extension cord out of the window to a portable generator during a power outage to power several lights and your refrigerator, this requires working in the dark or outside in a rain or snow storm. Wouldn't it be nicer to experience only a minor flicker of the lights when the utility grid fails and the backup generator starts?

Chapter XIV deals extensively with how to wire a battery powered inverter and solar photovoltaic system to power your home. Some readers may only want to use a manual start generator for their emergency or backup power system.

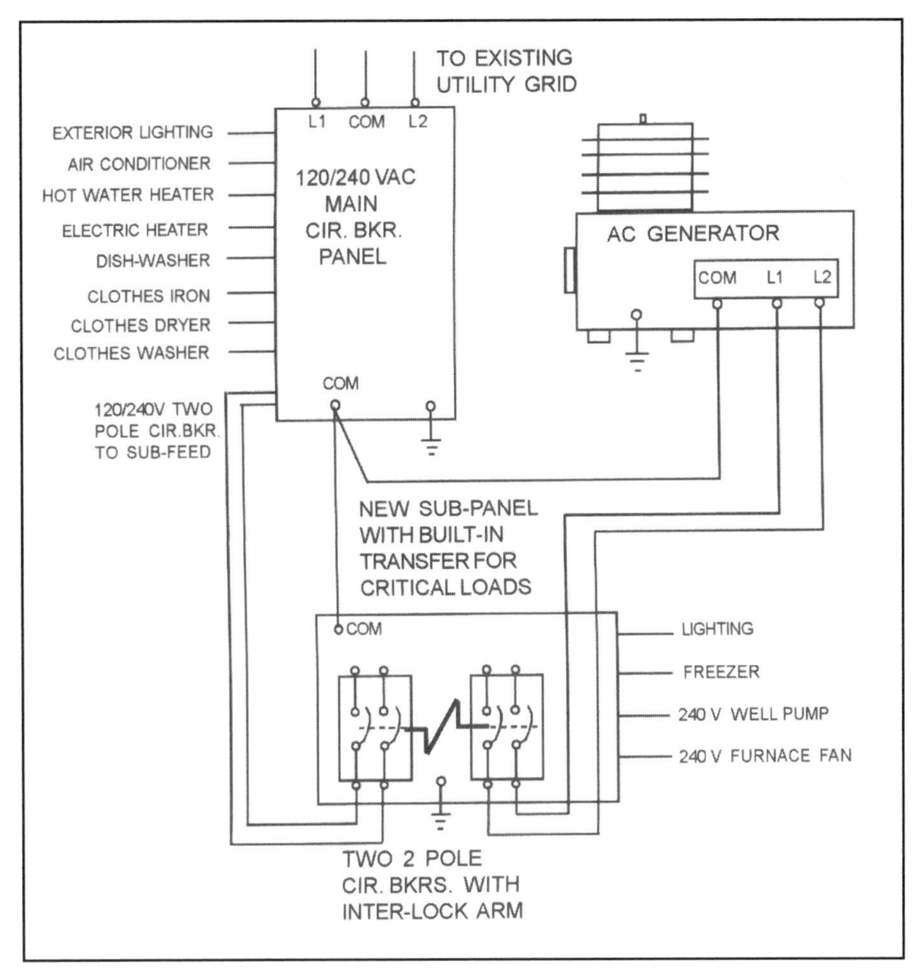

Figure 13 - 3 How to wire generator to serve critical loads using interlocked circuit breakers.

Mount the sub-breaker panel and transfer switch immediately next to your existing main circuit breaker panel. Try to mount at same elevation to allow easier access. If you will be using an inverter with your generator, it will probably include its own automatic internal transfer switch and the stand alone transfer switch shown will not be needed. However, since an inverter failure would keep the utility power from feeding through to the sub-panel, some designers still include a manual switch to bypass the inverter for servicing or repair.

The following discussion assumes you are installing a 4,000 to 6,500 watt 120/240 volt AC generator. These should be considered general instructions as all electrical work should be installed by licensed electricians, according to the National Electric Code. This may impose additional grounding or wiring

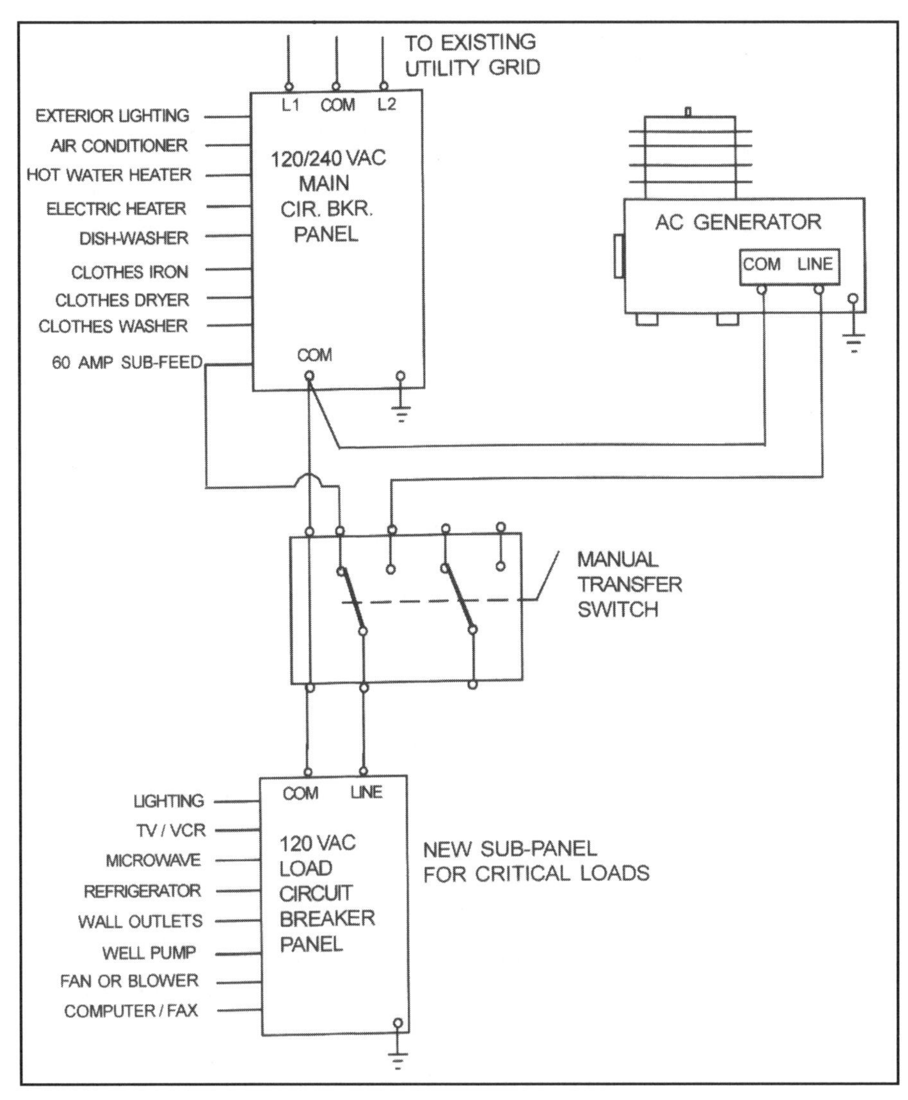

Figure 13 - 4 How to wire generator to serve critical loads sub-panel using transfer switch.

requirements, using only UL labeled wiring and hardware in order to pass inspection and insure the utmost safety. Larger generators and higher voltage loads will require higher current ratings for circuit breakers, an automatic transfer switch, and additional wiring.

Figure 13 - 5 Buss Jumper installed to convert 120/240 volt circuit breaker panel to 120 volt only. Jumper is larger black wire at top.

Have the electrician install a new circuit breaker in your existing house panel to feed the transfer switch as shown in Figure 13 - 3 and 13 - 4. Depending on generator size and loads, this circuit breaker will probably be in the 40 to 50 amp range. The electrician should next provide a new wire run from the transfer switch out to the generator location (preferably run underground).

If you are using the smaller sub-panel that includes a built-in double circuit breaker transfer switch, then you are finished. If you are using a separate transfer switch and circuit breaker panel, you will now need to connect the base leg of the transfer switch to the main buss bars in the new sub-breaker panel. Both buss bars can be tied together as shown in Figure 13 - 5 if you will not have any double pole breakers or 240 volt loads. This will give a total of eight single pole circuit breaker slots, as most sub-panels are designed with four circuit breakers on each buss bar of a 120/240 volt supply feed. If you want to power a 240 volt well pump or power tools, the generator, main, and sub-feed panel will need to be wired for 120/240 volt dual voltage service.

Finally, have the electrician remove all existing circuits presently connected to the main house panel which you want to power from the generator, and reconnect to circuit breakers in the new sub-breaker panel. Although you may want all lighting circuits on emergency power, be careful connecting wall outlet and appliance circuits. It is doubtful that you have over 4,000 watts of lighting in your entire house, but a portable heater you accidently left plugged into a wall outlet circuit that was re-routed and connected into the new sub-panel, could exceed the capacity of a generator also powering the lighting, well pump, and other appliances.

The electrician can advise you concerning how to balance the loads, generator grounding, and identify which circuits feed which outlets. If you have specific loads you want to insure stay on the generator supply, have your electrician install a new dedicated wall outlet at the appliance and run an isolated feed directly back to a separate breaker in the new sub-panel. This wiring technique may be required for the kitchen refrigerator, garage freezer, or audio/video entertainment center. Also remember that some gas kitchen stoves and gas water heaters will still require electricity to operate the electronic ignition pilot lights and temperature controls.

Do not plan to operate an electric hot water tank or air conditioner from an emergency generator unless you are planning a very large stationary power plant. You should be able to power a typical 1/4 to 1/2 HP fan motor in a central forced air gas furnace, or small circulating pump and controls for a hot water boiler heating system. A well pump, air handling unit fan, and a refrigerator or freezer operating together will exceed the motor starting capacity of most small 4 kW generators, so be sure only critical loads are connected or a larger generator is used.

When the rewiring is complete, all new wiring and switch controls should be permanently labeled. Any 120/240 volt sub-panels that have been wired for 120 volt only service should have labeling that clearly indicates no 240 volt circuit breakers can be installed. Emergencies are no time to guess what switch or circuit breaker feeds which loads.

CHAPTER XIV
BASIC WIRING SYSTEM LAYOUTS

In this chapter we will review how to assemble solar and backup power systems for residential and small business applications. I will start with a 12 volt DC only system usually found in small weekend use cabins or backup emergency lighting only applications. We will then review larger systems utilizing one or more AC inverters capable of powering all lighting and appliances in an off-grid home. The final example will be a small portable backup power system. For those individuals experienced with electrical wiring, these systems are no more complicated than connecting a component stereo system.

For those less experienced, please keep in mind the block wiring diagrams may not show all grounding connections, conduit, or wire sizes which must be installed per the National Electric Code. If you are not familiar with these requirements, I strongly suggest hiring a licensed electrician to connect the electric components for you.

When reviewing these diagrams, observe how both sources of power, the solar array and the battery, are wired to the switch side of the solar charger disconnect. This insures the fuse holder and solar controller are de-energized from both power feeds for operator safety while changing the fuse or servicing the controller. Also note the positive and negative wires to and from the solar array and battery bank are always on the opposite ends from each other. This creates equal length wiring paths and balances the voltage drops to each battery and each solar module.

The darker lines shown in these diagrams represent larger high current cables and the lighter lines represent smaller low current wiring. The actual wire sizes should be based on the equipment manufacturer's instructions; however, wire sizes still must be adjusted when extra long lengths are required, and note higher ambient temperatures reduce the design capacity of a given wire size.

WEEKEND CABIN - DC ONLY: The term "weekend cabin" refers to any application using only 12 volt DC lighting or small appliances in a building that will only be used on weekends or vacations. You can get by with a very small solar system since it will have a week or more to store up the energy that will be consumed on the weekend.

Although this system does not offer the flexibility of larger AC inverter based systems which can operate 120 volt appliances, the recreational and boating industry offer lighting fixtures, stereos, small televisions, VCR's, and small kitchen appliances designed to operate directly from any 12 volt battery.

Figure 14 - 1 shows how easy it can be to install a small solar photovoltaic array on the roof of a weekend retreat. A better mounting design would run the supports in a vertical orientation instead of horizontal as shown, to avoid blocking water and snow run off.

Figure 14 - 2 illustrates the suggested method to wire two 100 watt, 12 volt solar modules to two 6 volt T-105 golf cart batteries. Note that the 6 volt batteries are wired in series to create 12 volts in order to match the solar module's voltage.

Figure 14 - 1 Two roof mounted 100 watt PV modules for weekend residence.

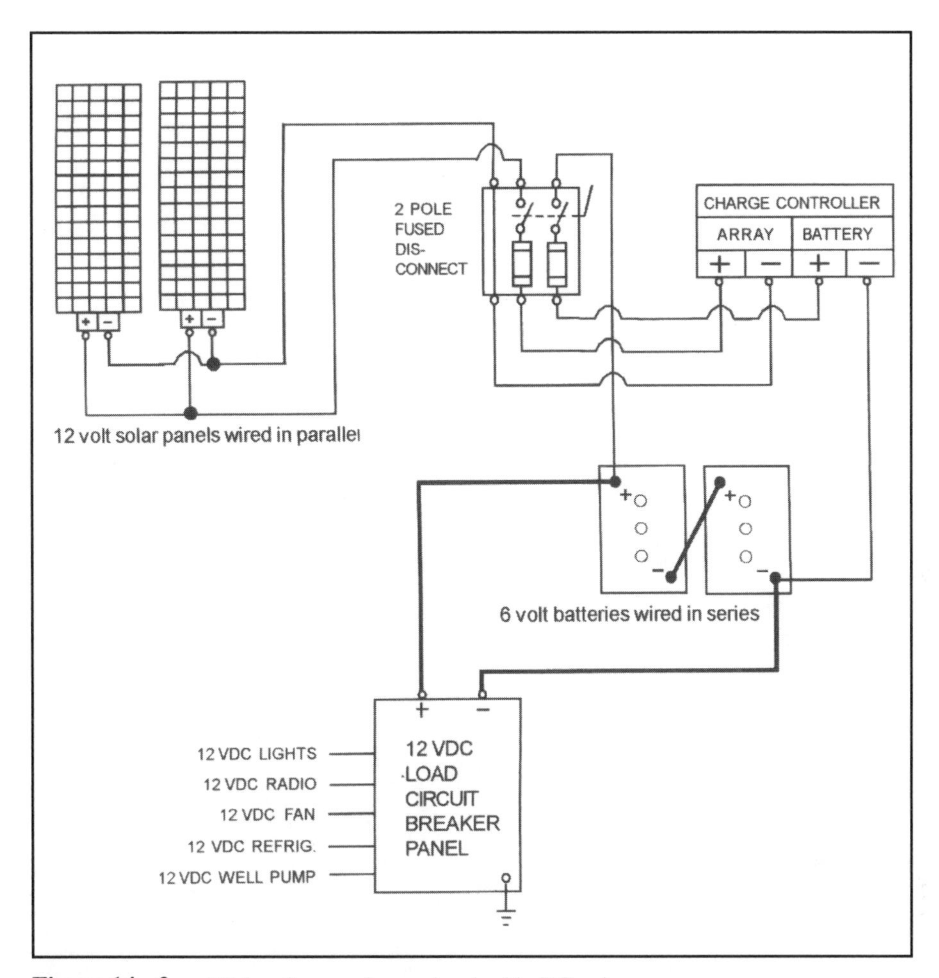

Figure 14 - 2 Wiring diagram for weekend cabin DC only.

WEEKEND CABIN - AC OPTION: By taking the basic design of the DC only system and adding additional solar modules and batteries, you can add a small inverter to power low wattage AC power tools and appliances. Due to the higher energy usage and limited solar array size, this system is still primarily intended for weekend only use.

The larger 200 to 250 amp DC circuit breaker shown on the following diagrams to protect the inverter can be ordered with its own separate enclosure as shown in Figure 14 - 3. Notice how two additional side mounted 15 to 60 amp DC circuit breakers can be added to feed DC lighting circuits, the solar charge controller, and other DC loads. A typical method to interconnect multiple L-16 deep cycle 6 volt batteries is shown in Figure 14 - 4.

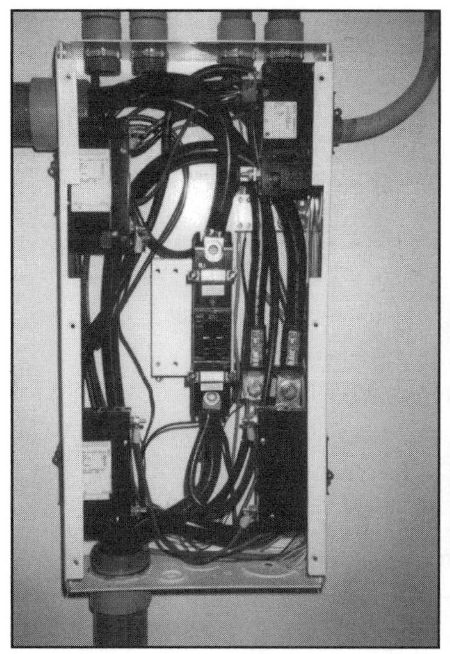

Figure 14 - 3 DC circuit breaker box protecting the inverter and other DC loads.

Figure 14 - 4 Typical battery interconnect wiring.

Figure 14 - 5 provides a modified version of the wiring diagram for the DC only weekend cabin shown in Figure 14-2. I have added two additional solar modules and two more batteries to increase total system capacity. To further increase the capacity of this system, 75 to 100 watt solar modules can be substituted for the 50 watt solar modules of the smaller system, but be sure the current capacity of the charge controller is increased accordingly.

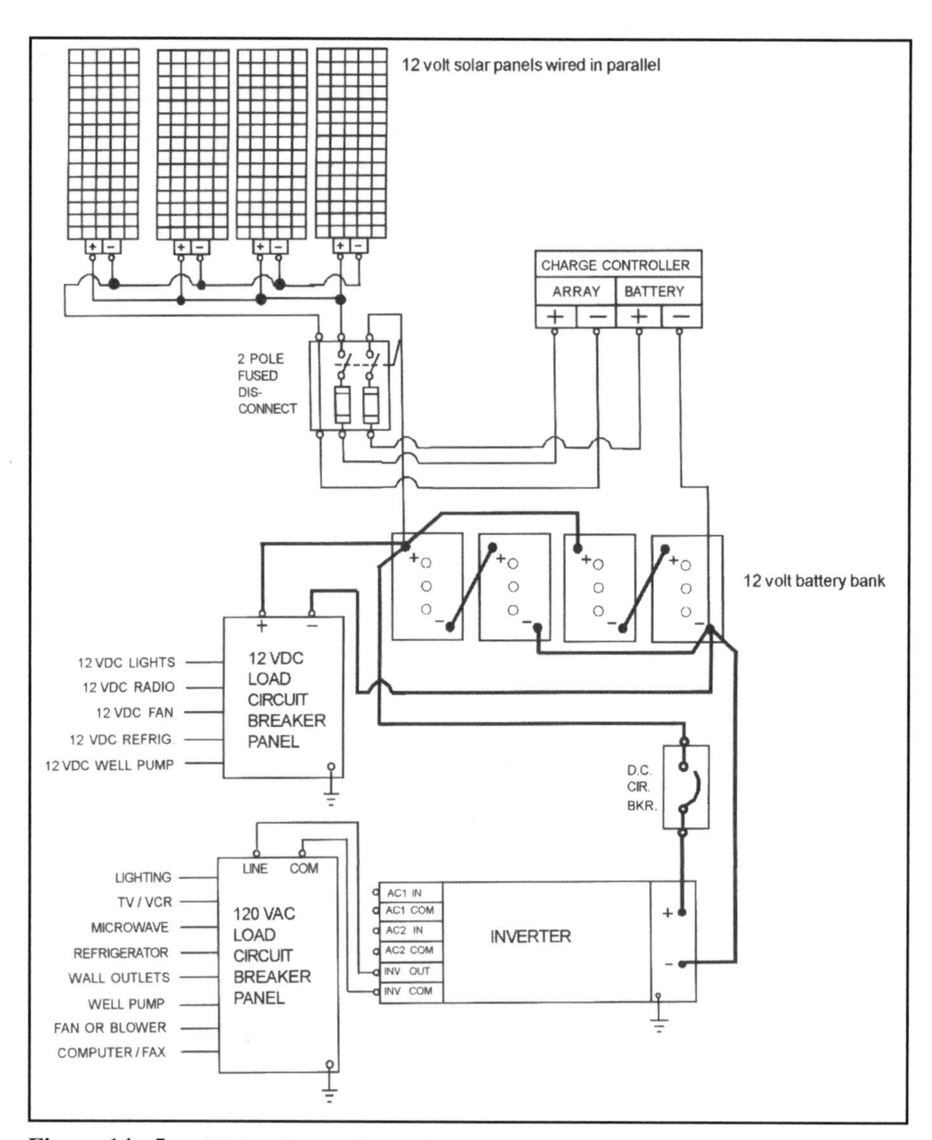

Figure 14 - 5 Wiring diagram for weekend cabin AC option.

TEMPORARY EMERGENCY BACKUP SYSTEM - 2500 WATTS:

This system is intended to keep emergency lighting, critical AC appliances, or office equipment operating in your home or office during a temporary power outage. The selection of components for the system assumes the power outage is temporary, and utilizes the utility grid to recharge the battery bank once power is restored. Since a generator is not used, when the battery charge is depleted, the system stops supplying power. This means the operating time is totally dependent on the number of batteries and the electrical consumption of the equipment being powered.

This system should be able to power the lighting in several rooms using compact fluorescent fixtures, an energy efficient computer, an answering machine, a fax machine, and a small color television or stereo, all at the same time for at least one full day before a total battery recharge is required. This system will not have the capacity to power large well pumps, power tools, or refrigerators.

When using an inverter with a utility grid system, it is recommended to install a new 120 volt AC circuit breaker panel like the 8 circuit panel shown in Figure 14 -6. All electrical loads which will be supplied from the inverter should be rerouted from the existing main house panel to this new panel.

Figure 14 - 6 Separate circuit breaker panel for loads on emergency circuit.

The 120 volt AC green neon pilot light shown at bottom center was added to the circuit breaker cover to indicate when power is available to power these loads. This will also make it much easier to see when the inverter has "tripped off" due to a low battery condition. Be sure to keep within the total output capacity of the inverter when planning which loads to supply.

Figure 14 - 7 provides a simple wiring diagram for this emergency backup power system. Note that solar modules and a solar charge controller are not required since the inverter is used to recharge the battery bank from the utility grid when power is restored.

The battery wiring shown is a 24 volt circuit since most larger inverters are designed for higher voltages to reduce the high current flows required with a 12 volt DC input. As stated in Chapter VIII, systems that only discharge batteries during power emergencies require batteries designed for this type of use. Sealed gel cell 6 volt batteries make a good choice for smaller emergency backup only systems, and glass case 2 volt cells are excellent for larger systems. Deep discharge liquid electrolyte batteries like the T-105 and L-16 will not reach their expected life with the very infrequent discharge cycle of this emergency standby aplication.

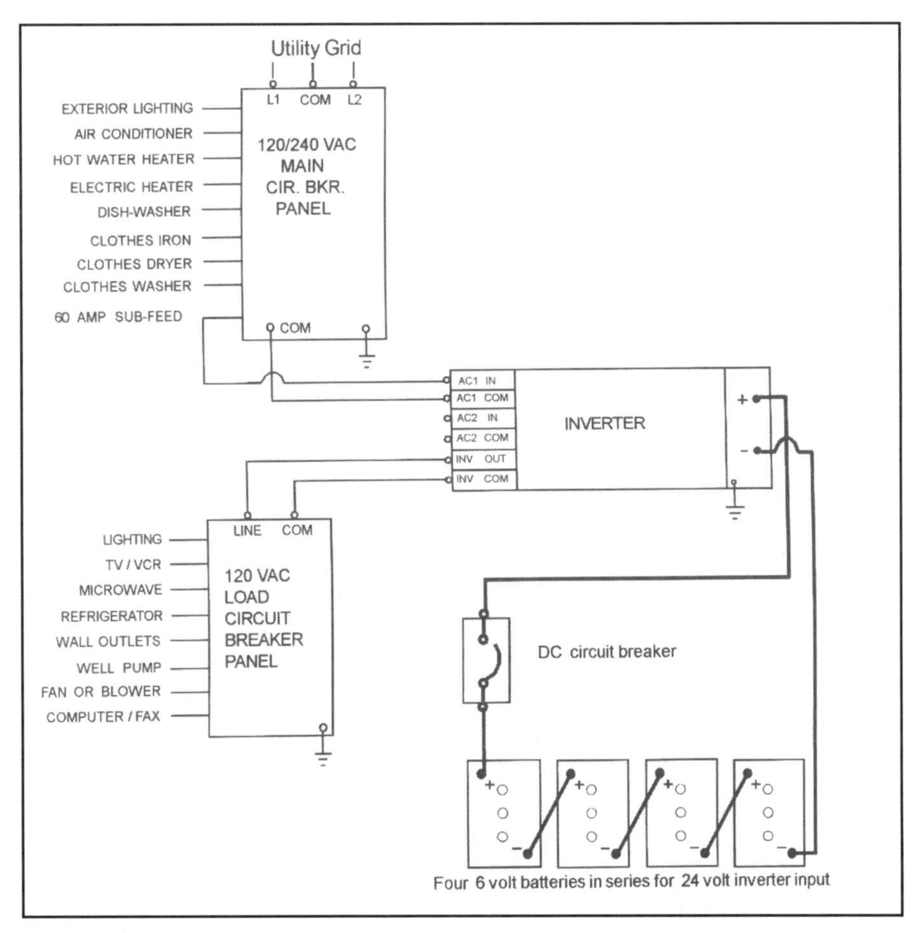

Figure 14 - 7 Simple wiring diagram for emergency backup power system.

BACKUP POWER SYSTEM - 4,000 WATTS: This system is very similar in the wiring layout for the 2,500 watt system, but adds a backup generator to recharge the batteries during extended power outages. It is still intended for emergency use applications utilizing the utility grid to recharge the battery bank once power is restored. This system is also very popular for homes or small businesses in rural areas with frequent power outages, or countries using rolling blackouts each day to limit the loads on hydroelectric plants during the dry season. Since the inverter includes a high capacity battery charger, the battery can be recharged in 4 to 6 hours. When planning a system that will use a generator to recharge the battery bank, the higher amp output battery charger is worth the added cost as long as the charger does not exceed the load capacity of the generator. For example, a quality 40 amp charger will cost two to three times the cost of a 10 amp charger, but your generator will operate only 1/4 as long per charge cycle saving fuel and generator life. Your neighbors will appreciate this also.

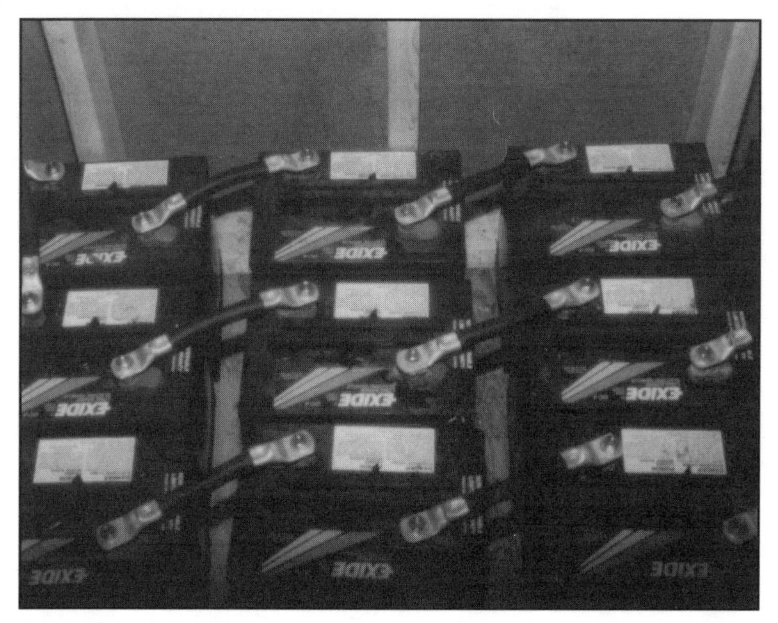

Figure 14 - 8 Six volt golf cart batteries wired in series - parallel for 24 volt.

Figure 14 - 8 provides a partial view of a 24 volt battery bank consisting of three rows of four 6 volt golf cart batteries wired in series - parallel. When planning your system layout, locate the inverter, circuit breakers and related electrical equipment together for easy access as shown in 14 - 9. Note the use of conduit for all wiring between components.

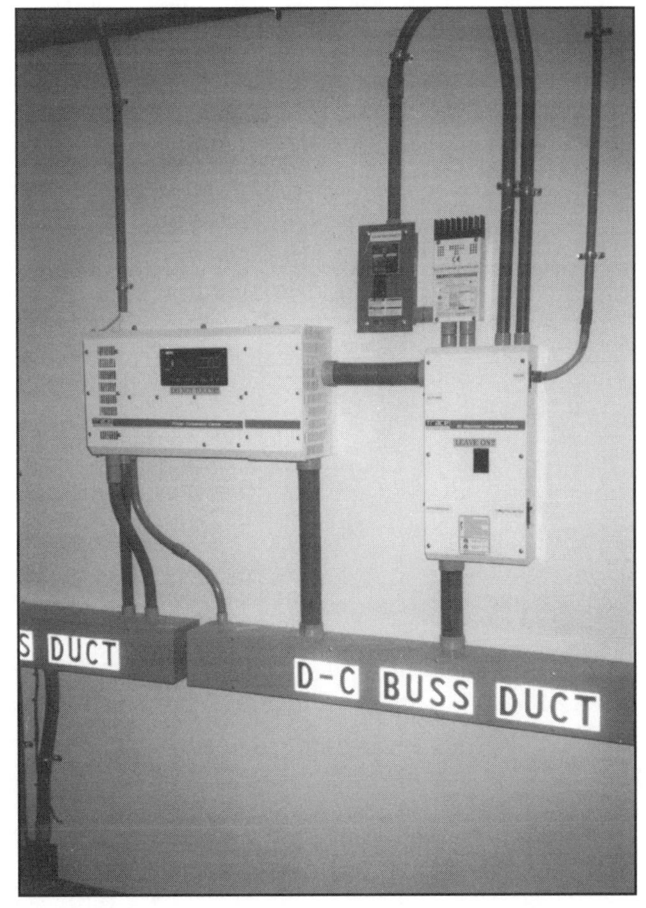

Figure 14 - 9 Surface mounted inverter and circuit breakers using exposed wiring trays and conduit.

Figure 14 - 10 shows how to add a generator to the system described in Figure 14 - 7. To reduce generator run time, I have added additional batteries.

To further increase battery capacity, the larger L-16 deep discharge 6 volt batteries can be substituted for 6 volt golf cart batteries. To reduce the higher currents flowing through the inverter power section and the inverter to battery cables, the batteries are wired in a 24 volt arrangement, and a 24 volt input inverter is substituted for a 12 volt input inverter.

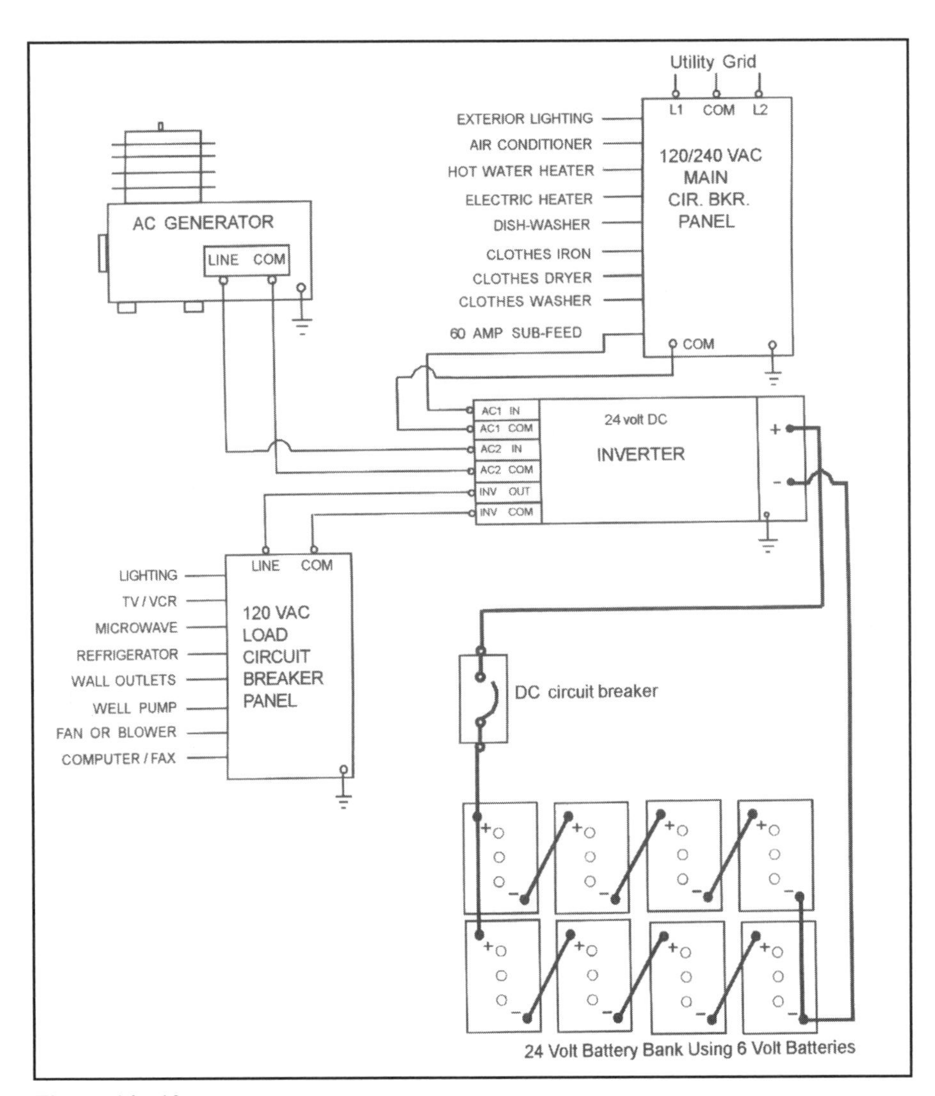

Figure 14 - 10 Wiring diagram adding generator.

OFF-GRID SOLAR SYSTEM - 4,000 WATTS: A solar system this size could still be connected to the utility grid, but the utility is used primarily for temporary peak loads or just to power an air conditioner or large occasional use appliance or shop tools. Due to the higher capacity of the inverter, battery bank, and solar array, the homeowner has the peace of mind that the utility grid is not their primary source of power, and power outages are no longer a concern.

The larger 240 volt AC appliances like shop tools, a cloths dryer, or air conditioner are wired directly to the utility grid main house panel, while all other more critical electrical loads including lighting, wall outlets, kitchen appliances, and entertainment equipment are connected to a sub-panel powered from an inverter and solar charged batteries. The homeowner still enjoys the conveniences and comfort their larger 240 volt AC appliances provide, yet these larger appliances are not critical to operate during week long power outages. A wood stove, and a gas or propane kitchen stove and water heater can also lessen the inconvenience of a long term loss of utility power.

To maximize the solar performance of smaller systems, it is helpful to mount the modules on one or more hinged frames as shown in Figure 14 - 11 which allows spring and fall seasonal array tilt adjustments. In this example, two groups of eight modules are mounted on separately hinged aluminum frames. Using a separate mounting frame for each group of six to eight modules makes it easier for one person to adjust the tilt angle. Although a dome home can be a design challenge to install a straight row of solar modules, Figure 14 - 11 shows how a south facing deck can be used to mount a small off-grid array.

Figure 14 - 11 Rural off-grid 16 module deck mounted array.

This system wiring is shown in Figure 14 - 12 and includes a much larger solar array and battery bank. On systems this size, several manufacturers offer a wall mounted power module that includes an inverter pre-wired to the main DC circuit breaker and solar controller, with extra low voltage circuit breakers for emergency DC lighting and other DC loads. (See Figure 14 - 3). Although heavier to transport and mount, this greatly simplifies the field wiring. The inverter and batteries are usually wired for 24 or 48 volt DC operation.

The heavier L-16 style 6 volt battery makes an excellent choice for this system as it will experience a charge and discharge cycle almost every day. Note how multiple batteries are wired in a series - parallel arrangement to increase battery bank voltage anc capacity. Also note how the positive and negative leads are located at opposite ends of the battery bank to maintain a uniform voltage drop to each individual battery.

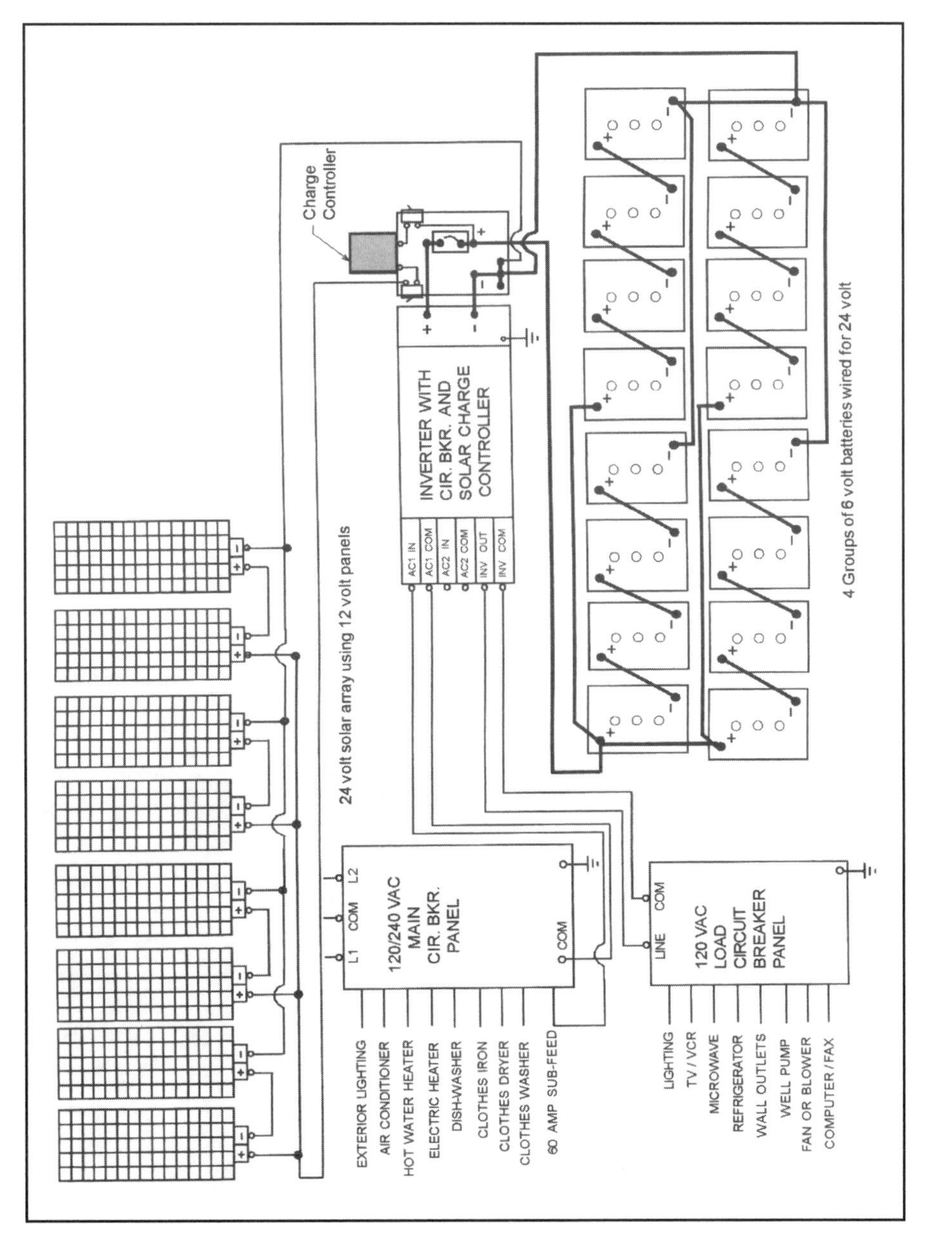

Figure 14 - 12 400 watt off-grid wiring diagram.

OFF-GRID SOLAR SYSTEM - 8,000 WATTS: Due to the substantial investment required for systems in this size range, these larger 120/240 volt dual AC inverter systems with solar and generator charged battery banks are usually found in larger family residences located in areas of the country not served by the utility grid. Homeowners wanting to build in these beautiful, but rugged parts of the country are finding out that these fairly expensive systems still cost less than the fees the power company wants to charge to extend utility power lines to a remote site.

To reduce the large number of solar modules and batteries required, these systems usually include a propane fueled generator in the 7 to 10 kW range. The primary purpose of the generator is to recharge the battery bank during long periods of cloudy weather, and to power intermittent large electric loads like shop power tools or a clothes washer. For example, when washing clothes, it is not unusual to also operate a clothes dryer, well pump, and clothes iron, all at the same time. Although this large dual inverter system could handle these combined loads, the efficiency losses in charging the batteries, and the added efficiency losses through the inverter to power these very large loads makes it more cost effective to only operate these larger loads while operating the generator to recharge the batteries.

Figure 14 - 13 Large 2.4 kW ground mount array.

Figure 14 - 13 proves not all photovoltaic arrays need to be roof mounted. This 2.4 kW array operates at 48 volts DC to reduce electric line losses caused by the longer cable distance to the house. Note the use of landscaping and screening to reduce the visual impact of this large array structure. The adjustable tilt aluminum mounting rack is bolted to 8" diameter concrete piers formed using cardboard tubes.

Large off-grid systems usually have inverters and battery banks wired for 24 or 48 volt DC operation to reduce high current flows. As the voltage of the solar array and battery bank are increased, wiring and power components can be reduced in size and cost.

Due to the use of multiple inverters, multiple solar module sub-arrays, and higher battery voltages, the electrical system layout can become complex on these larger systems as shown in Figure 14 - 14 and should be left to a solar professional.

Figure 14 - 14 Larger off-grid solar and generator control center.

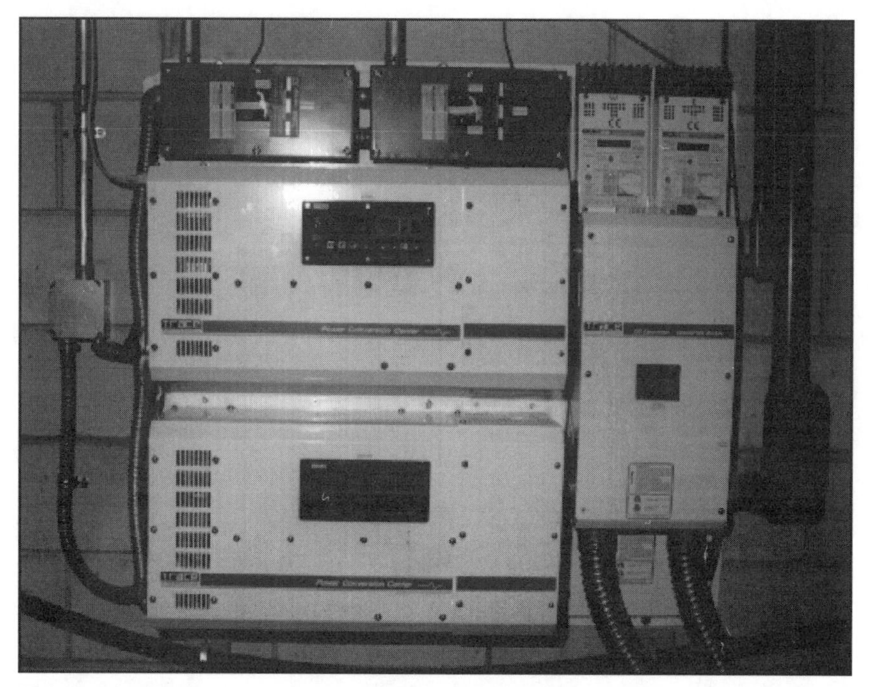

Figure 14 - 15 Trace Engineering dual inverter pre-wired power panel.

Figure 14 - 15 shows an example of a complete backup power system that comes pre-wired with the inverters, bypass switches, and low voltage DC circuit breakers, mounted on a single back plate. All that is required to install this system is connecting the battery cables to the batteries and the bypass switches to the existing main house circuit breaker panel. If the system will include a solar array, the factory wired power panel can be supplied with one or more solar photovoltaic charge controllers and associated DC circuit breakers also pre-wired which only requires the connection of the array feed.

More and more manufacturers are now supplying their inverters already wired to the electrical switchgear which greatly simplifies field wiring. Another advantage of using these pre-wired panels is the assembly and all wiring components are UL labeled which can make electrical inspections much easier.

GRID TIED PHOTOVOLTAIC SYSTEM: In reality, most solar powered homes will still use the utility grid for powering the occasional large load or for backup. What most solar professionals actually mean by grid tied is a photovoltaic array wired directly through a DC to AC inverter into the utility grid using an electric meter that can measure electric flows in both directions.

These grid tied systems are usually in the 2 to 6 kW size range and are installed in locations where the local utility charges very high summer afternoon peak kW demand charges. By off-setting electric demand during these peak periods, significant dollar savings can be realized without regard to energy storage. Since the utility grid becomes an infinite storage system, these systems do not need any battery storage. After the peak period has passed, the home or business consumes power back from the grid at substantially discounted rates. Many utilities are now required to purchase this "green power" from independent power producers due to recent state legislation. Buy back rates in excess of $0.50 per kWh are not unusual, which is substantially higher than standard rates.

Prior to recent changes to state utility regulation, it was almost always cost prohibitive to use a solar or wind alternative energy system to feed electricity back into the grid. Many utility companies required the homeowner to install very expensive safety interlock switchgear and carry very high liability insurance to protect the utility company, not the homeowner, from all future liability. As more and more people choose to purchase "green power" as part of the deregulation of the electric industry, more reasonable heads will prevail.

To reduce the confusion of each state and each utility company requiring different grid interconnect requirements, the IEEE is sponsoring a national grid interconnect standard, but at this date it has not been adopted by all utilities.

Figure 14 -16 provides a close-up view of a grid tied array with the mounting feet incorporated into the module frames. Note how the modules are directly attached to a shingle roof using lag screws at the interlocking frame sides.

Figure 14 - 16 Grid tied solar modules designed for direct roof mounting.

Thin film solar modules are usually used in direct roof mounting applications as they can better withstand the higher operating temperatures near the roof's surface than polycrystalline module designs. This allows direct module to roof attachment without a separate stand off mounting structure. Thin film technology solar modules have a much lower watt per square foot output than polycrystalline modules, so much more roof area will be required for solar arrays using these modules.

Due to the higher voltages and safety issues involved, these systems should be installed by licensed electricians. The thin film solar modules shown in Figure 14 - 16 are designed to provide 120 volts DC for each pair wired in series. Higher voltage modules allow using much smaller interconnect wiring. Some solar modules are available with pre-wired "pigtails" at each end or at opposite sides which allow easier series wiring connection.

Many people installing grid tied solar photovoltaic systems do not always realize their solar electricity will stop anytime the utility grid fails since the inverter cannot stabilize the system voltage without the grid serving as the missing battery bank.

An example of the new Trace direct line tie inverter is shown in Figure 14 - 17 mounted on an exterior wall directly beside the utility meter. These inverters are designed for a direct grid tied application without battery storage. All electrical disconnects, fuses, and controls come pre-wired in a weatherproof enclosure which allows quick installation.

Most grid tied inverters are designed to be located outdoors or in an accessible garage to allow easier access to utility company personnel when necessary to isolate system during line repair work.

Figure 14 - 17
Inverter designed for a grid tied application without battery storage.

The advantages of not needing space for a battery bank and no battery maintenance requirements may be lost if a perfectly functioning solar array cannot supply your home with power during an extended electrical outage. On the other hand, many grid tied solar arrays are installed only to reduce total yearly grid electrical usage and not to provide emergency power. An occasional power outage may not be objectionable in this application.

Figure 14 - 18 is a basic wiring layout for a grid tied solar system. This simplified block diagram does not show all safety interlock controls and system disconnects that may be required by the local electric utility to allow de-energizing the system by their personnel at any time of the day or night. Note the use of a circuit breaker in the main house panel to allow the inverter to "back feed" power back through the house panel and out to the grid.

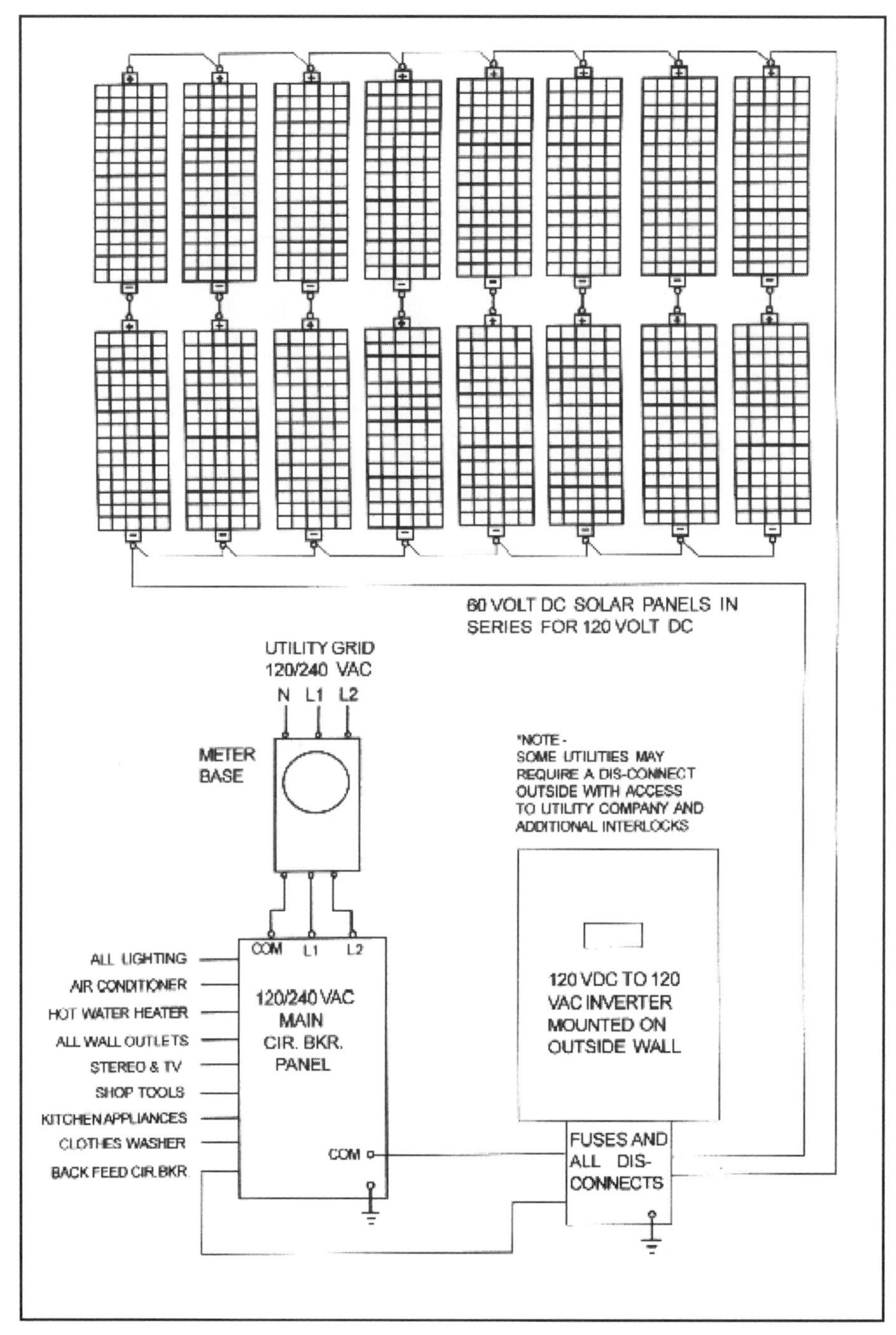

Figure 14 - 18 Grid tied solar array without battery bank.

WIND CHARGING OPTION: All of the previously described solar and backup power systems can easily have a wind turbine generator added to provide additional battery charging capacity. Chapter VII described the site selection issues with wind generators so I will now address several wiring issues that are very important but usually overlooked.

Due to the very high cost of solar photovoltaic modules, it is unusual to have an oversized solar array. In most cases, the reverse is true and the batteries are always able to absorb all of the energy delivered. Even though most residential size wind generators only deliver a few hundred watts of power, there are many coastline locations where this small charging output is almost continuous. For this situation, it is not uncommon for the wind generator to fully recharge a battery bank and still have hours or days more of windy weather. However, unlike photovoltaic modules, a wind turbine needs an electrical load to prevent over-speed damage.

Rather than waste this unused battery charging energy, most battery chargers designed to operate from a wind generator will also include a "diverter" circuit. Once the batteries have been fully charged, the full output of the wind generator is automatically switched to a second set of contacts to power an auxiliary DC electric load. The most common application for this excess electricity is hot water heating. Several different types of 12 and 24 volt DC heating elements are available to replace the lower 240 volt AC element in a conventional electric hot water tank.

Since the generation of electricity absorbs the energy of the wind, disconnecting the wind generator from its electric load removes the loading on the wind turbine blades which can result in very high rotation speeds and noise. This is why many wind turbine generators also include a "stop" switch which keeps the blades from turning when power is not needed, or for operator safety during servicing.

Many models of wind generators, like the Model AIR from Southwest Windpower Incorporated shown in Figure 14 - 19, have electrical characteristics designed into their charging circuit to duplicate the charging voltage curve of a photovoltaic solar array.

Figure 14 - 19 Wind generator mounted above solar array.

This allows using a standard solar charge controller for the wind generator circuit. This can simplify installation, but locations with fairly good wind availability will probably use a charger designed with the extra diverter circuit as shown on the wiring diagram in Figure 14 - 20.

A residential water wheel turbine generates electricity by being piped to a water source located at a higher elevation. A water wheel turbine generator can be wired into the system using the same method as used for the wind generator. Like the wind turbine, it also needs a constant electrical load to prevent over-speed damage.

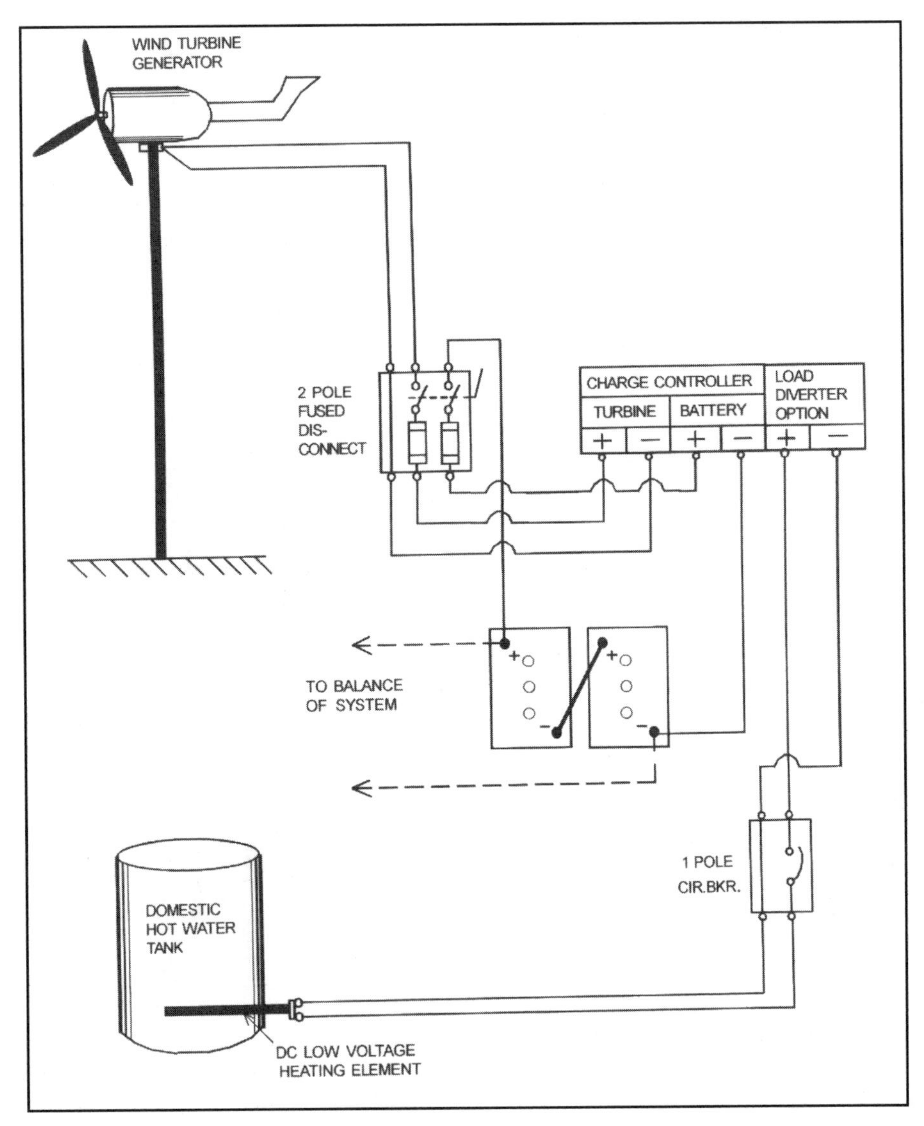

Figure 14 - 20 Wind turbine generator option.

SOLAR POWERED FARM WELL PUMP: There are several deep well 12 and 24 volt pumps available that are designed to slowly fill above ground stock watering tubs and non- pressurized domestic water storage tanks.

Figure 14 - 21 shows a diaphragm type 24 volt DC submersible deep well pump. This pump utilizes a rubber diaphragm that is pulled up and down by an electric coil to create the pumping action. Although very energy efficient and capable of high head pressures, this style of pump has lower flow rates compared to larger motor driven centrifugal pumps and requires periodic replacement of the diaphragm.

Figure 14 - 21 SolarJack 24 volt DC submersible well pump. Note size of ink pen.

A 24 volt DC centrifugal pump can be seen in Figure 5 - 2 in Chapter V mounted on the floor next to an indoor storage tank. This type of pump would work well with this system and would provide higher flow rates if it could be located near the water source and protected from freezing.

Since these systems are usually located out in a remote field and away from occupied dwellings, it is helpful to design a system that does not need a battery bank which requires maintenance and protection from the weather. Without a storage battery however, the pump will stall out and stop pumping as the voltage from the solar module fluctuates from passing clouds. Since this can overheat and damage most pumps, it is important to use a pump controller and not wire the pump directly to the solar modules. These controllers are designed to constantly adjust the voltage and current being supplied to the pump for maximum performance, and cut off the power when it is too low for safe operation.

These pump controllers also include wiring connections for high level and low level limit switches for more complex control arrangements.

Although not required, these systems usually have a pole mount solar array instead of ground mount to avoid damage from livestock or the shade of any nearby trees. The wiring for this system is very simple as shown in Figure 14-22.

Most pump manufacturers can also provide a pump controller designed to maximize their pump's performance in this changing voltage application.

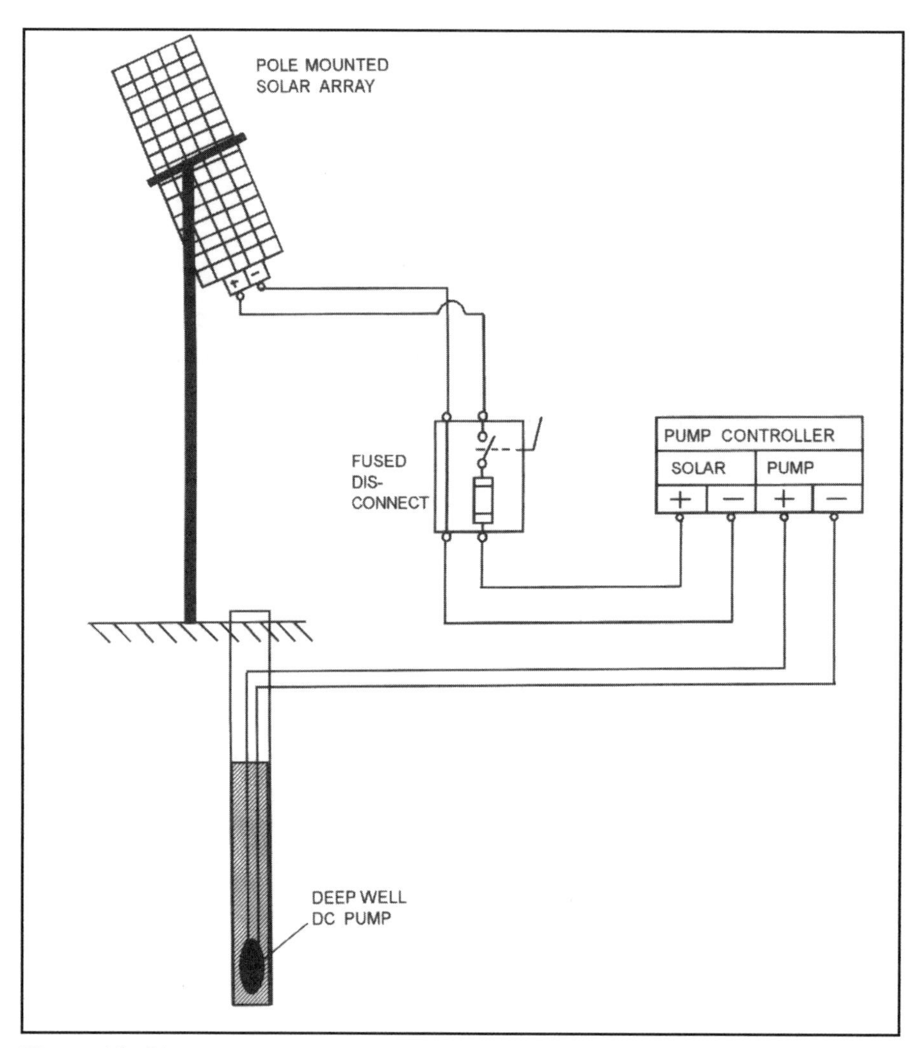

Figure 14 - 22 Solar powered farm well pump.

PORTABLE BACKUP POWER SYSTEM - 500 WATTS: This is a great weekend project for anyone wanting to build their own portable backup power system using parts easily obtained locally. Since the system is self contained, it can be used to power small hand tools or appliances at the job or camp site as shown in Figure 14 - 23 or power a home office during temporary storm power outages as shown in Figure 14 - 24.

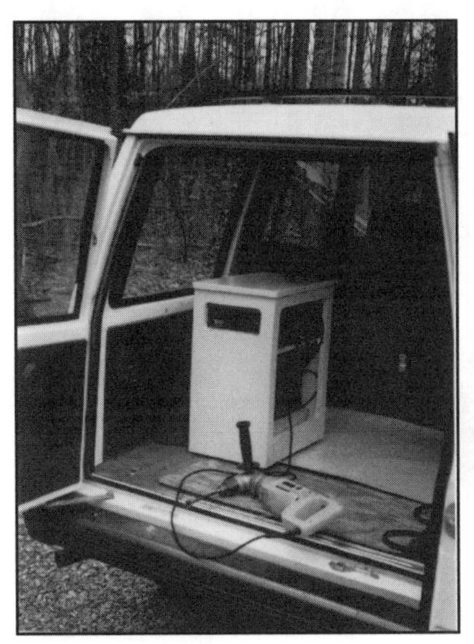

Figure 14 - 23
Portable power system at job site.

Figure 14 - 24
Portable power system in the home office.

Almost all RV and boating supply retail stores stock small inverters, battery chargers, and deep cycle batteries. In addition, they also stock several sizes of high current capacity battery cables and connectors. To keep weight to a minimum, I suggest using two six volt deep cycle golf cart batteries or one 12 volt marine battery, a 500 watt 12 volt inverter that includes a standard duplex AC outlet, and a quality 10 amp battery charger that includes both bulk and float charging cycles.

Figure 14 - 25 shows the cabinet fabricated from 3/4" plywood with openings allowing access to the charger controls, inverter, and batteries. When planning the cabinet dimensions, you may want to incorporate other features such as a hinged top, swivel rollers, and lifting handles.

Keep all cables as short as possible. Since the battery charger cables may come with large alligator clips, these need to be removed to allow direct wiring to the battery terminals. Be sure to use red tape to mark which lead is positive before cutting off the clips.

Figure 14 - 25 Interior view of portable power system using single 12 volt marine battery.

Many portable inverters in this size range are available with useful features including battery state of charge indicator, automatic low battery disconnect, and a remote start/stop switch. Figure 14 - 26 provides wiring diagrams for a system using two 6 volt golf cart batteries, and a system using one 12 volt deep cycle marine battery.

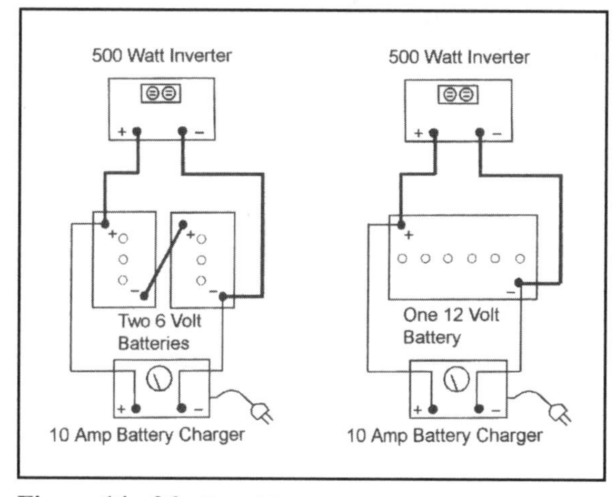

Figure 14 - 26 Portable power system wiring diagram showing both battery configurations.

The system pictured has a single 12 volt deep cycle marine battery and was able to power a Pentium color laptop computer and two office lighting fixtures for four hours during tests. Two or more 13 watt compact fluorescent portable "drop lights" with long extension cord plugs make excellent emergency lighting for use with this system.

If the portable power system will stay inside during battery charging, you may want to use the slightly more expensive gel cell batteries which do not vent hydrogen gas during charging or spill acid if turned over. When selecting the battery charger, remember that gel cell batteries and wet cell batteries require different charging voltages.

CHAPTER XV
WIRE SIZING

Although I promised to keep this text as non-technical as possible, there are many readers who have requested at least some basic "rule of thumb" design guidelines for preliminary wire sizing. In order to provide these guidelines, remember it takes the entire 1016 page National Electric Code to address all possible wiring situations, and these cannot be summarized in a few paragraphs. Always refer to the NEC Handbook before making your final wire selection.

Figure 15 - 1 shows several types of wire that are used in backup power and solar photovoltaic systems. The wire to the left is a #10 type THHW along with a crimp-on fork style wire terminal. THHW type wire must be installed in conduit. The center wire is a #8 type USE-2 underground wire with a crimp-on ring terminal. USE-2 wire can be direct buried without conduit. The heavy cable on the right is a #4/0 battery cable along with a copper connector lug. The dime in the picture is for size reference.

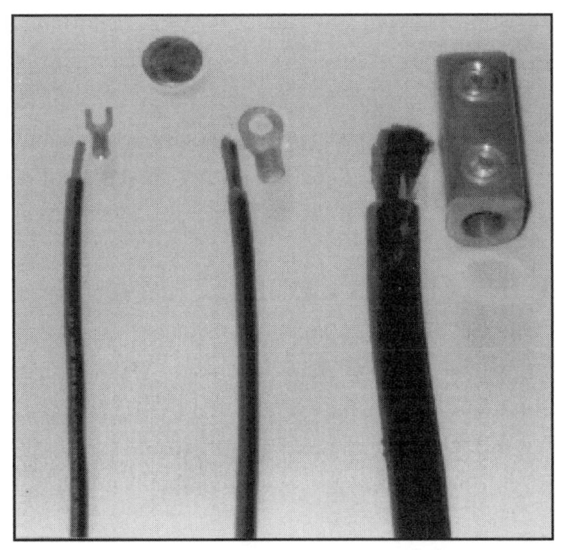

Figure 15 - 1 Typical wire types used in solar and backup wiring.

When selecting a wire to connect two points in any electric circuit, we need to know more than the current and voltage. There are many variables that can affect our wire selection including: will it be located indoor or outdoor, is it in conduit, is it exposed to sunlight or water, is it above or below ground, do you need a single or multiple conductor within the same cable, and the ambient temperature of the proposed location. Before providing a wire sizing guide, we need to briefly review how these items may apply to your situation, and how these could change your final wire size selection.

WIRE TERMINALS: Wire sizes above #8 will usually require a crimped and soldered copper terminal for bolt on connection. Be sure to use a ratchet type commercial crimping tool that will not release until fully compressed to install the wire terminals. Many professionally made cables are also soldered after crimping. The exception to this is several brands of large DC rated circuit breakers designed to accept bare wire ends up to 4/0 in size. All crimped-on wire terminals should be tinned copper, not plated steel or aluminum. When connecting very large battery cables to fuse blocks or buss bars with bolts, be sure that the surfaces of each cable terminal makes direct contact with the other, and is not separated by washers. Stainless steel washers are poor conductors, and at high currents can cause terminal over heating if improperly installed. This has been known to cause premature failure of these fuses.

WIRE COLOR CODES: The NEC specifies that black is "hot," white is "neutral," and bare copper or green is ground for all AC and DC wiring. Since most large cables used in DC solar and battery wiring have black insulation, the positive conductor is sometimes indicated by red terminals. When using a cable that does not follow this color code, be sure to add electrical tape having the designated color for at least 3" of length at each end.

WIRE SIZE VERSES RESISTANCE LOSS: All wire has a resistance that will cause a voltage drop. Just because we connect one end of a wire to a 12 volt battery does not mean we will get 12 volts of power out the other end. As the wire run increases in length, the resistance can increase substantially for higher current flows, causing an unacceptable loss of voltage. This loss is more noticeable in low voltage applications as there is a lower voltage to start with.

Most of the wire sizing tables you will find in other texts and in the electrician's tool box calculate wire sizes on an allowable 5% voltage drop. This is a very reasonable value for a conventional residence and acceptable to the NEC, but is entirely unacceptable for an off-grid or super efficient solar home. All wiring for our work should be based on a 2% voltage drop, and this includes wall outlet circuits, switch circuits, lighting circuits, and especially any battery, inverter, and solar array wiring. This usually results in a one size increase in wire size with only a slight increase in cost. Longer wire runs may require more than one size increase. In low voltage DC battery and inverter wiring involving very high currents, 15 feet is considered a very long wire run, whereas in higher voltage AC wiring, it may take over 100 feet before there will be any significant increase in voltage drop. All wiring and wiring terminals should be copper, and larger wire sizes should use stranded, not solid copper conductors to make bending the larger cable easier. Due to the lower conductivity and increased problems of terminal oxidation, aluminum wire and connectors should never be used in off-grid or solar homes.

WIRE SIZE VERSES TEMPERATURE: All wire manufactured to meet the National Electric Code and Underwriter Laboratories (UL) requirements must be labeled with an identification code that is used to classify the wire insulation. This also indicates the highest temperature this insulation can safely withstand without derating the wire current capacity. This is a real concern when wiring to and between solar modules, as this wire must be suitable for wet locations even if installed in conduit, be sunlight resistant if not installed in conduit, and have a very high temperature rating since the wire will be connected to wiring terminals mounted on the back of a very hot solar module. Do not use any wire to interconnect solar arrays with an insulation rated below 75°C (167°F), and be sure to derate the current capacity of the wire for higher ambient temperature as indicated in TABLE 15 - 1.

If solar interconnect wiring will not be installed in conduit, most installers use single conductor type USE - 2 which has a 90°C (194°F) insulation that is both water and sunlight resistant.

WIRE SIZE VERSES CONDUIT: Conduit is available in rigid metal, rigid plastic, flexible metal, and flexible plastic materials to enclose multiple and single conductor wires going to separate components. This allows using dimensionally smaller wire having less expensive insulation since the mechanical and weather protection is provided by the conduit. Conduit allows solar modules to be interconnected with THWN-2 grade wire which is much easier to find and less expensive than type USE - 2 wire. When installed in conduit, the wire will have a lower maximum allowable current rating, and even lower rating when multiple wires are routed in the same conduit. This is due to the higher ambient temperatures and the lack of the heat sink effect for a direct bury or open air installation. The NEC addresses these issues in detail, including temperature adjustment factors.

WIRE SIZE VERSES SAFETY: It is obvious that a wire must be sized to carry the anticipated current flow, but what is the anticipated current flow, and how much extra should we design into the wire selection as a safety margin? All wire design tables are based on the assumption that wire and electrical panels should not be subjected to their maximum current ratings 100% of the time, so allowance is made to derate the wire by multiplying the expected load current by 125%. The exception to this rule is the wire between a generator and the first circuit breaker. This generator feed can be sized for 115% of the generator's rated current output capacity.

When sizing the wire for a solar array, most people would read the current rating printed on the solar module name tag, and add up the "strings" of solar modules connected in parallel to obtain the total current the wire must carry; however, this is wrong and will produce unsafe wire sizes. The NEC requires using the "short circuit" current rating of a solar module for wire sizing, and this is much higher than the at load current rating.

In addition, unlike other electrical components with a known maximum current, a solar photovoltaic module can actually produce higher than nameplate current flows on partially cloudy or snowy days, since reflected sunlight from these surfaces causes even higher current flows. A second 125% factor is added per UL requirements to the wiring design factor which results in a 1.56 total wiring safety formula (1.25 x 1.25 = 1.56) which applies to any wire connecting a solar module or solar array to each other or to a charge controller. All other non-solar circuit wiring will have only the single load factor applied.

WIRE SIZE VERSES CIRCUIT BREAKER AND FUSE SIZE: Selecting a wire size based on the anticipated normal current flow still does not guarantee that the wire will not be subjected to higher than safe levels of current flows if equipment problems or short circuits occur. This is the purpose of the circuit breaker or current limiting fuse. The circuit breaker senses when the current passing through the wiring circuit exceeds the maximum current allowed for this wire size and interrupts the circuit. Of course the wire must temporarily carry this higher current flow without melting or overheating or the higher current flow would not trip the breaker before the wire failed. For this reason, the circuit breaker or fuse must always be sized with a trip setting below the maximum current capacity of the wire being protected, and a minimum of 125% of the maximum short circuit current.

Many electrical fires are started by overloaded circuits causing current flows greater than a wire can safely carry, and care must be taken when selecting a wire size. Many people assume because a wire is carrying only 12 or 24 volts DC there will less fire danger than with 120 volts AC. Keep in mind it is the current flow, not the voltage, that overheats a wire and eventually leads to melted components and failure. If you have any doubt as to this advise, ask any mechanic what happened when they accidently dropped a screwdriver across both terminals of a 12 volt car battery. DC rated circuit breakers for solar and battery bank applications are available in 5, 10, 15, 20, 30, 60, 100, 175, and 250 amp sizes. If your circuit is between circuit breaker sizes, use the smaller breaker or increase the wire size to allow using the next larger circuit breaker, but do not exceed the current limit set by the manufacturer for the component being connected.

FUSES VERSES CIRCUIT BREAKERS: This age old question is like asking "which is better, Ford or Chevrolet?" For all practical purposes both the fuse and circuit breaker do the same job, but each have different advantages and disadvantages. Fuses generally can interrupt a short circuit faster than an equal sized circuit breaker, but the fuse is destroyed in the process and must be replaced. AC circuits have a constantly reversing current flow which makes it much easier to interrupt the current at the instant it reverses and passes the zero voltage point 120 times each second (2 x 60 cycles). A DC current flow is constant and once started, is very hard to stop. An AC rated fuse or circuit breaker used in a DC circuit will open at the intended current limit; however, the DC current will usually just arc across the AC contacts and continue to flow. DC rated fuses, circuit breakers, and wall switches are much heavier constructed and are designed to extinguish this internal contact arc. Never use AC rated devices in DC circuits unless they are also DC rated.

Circuit breakers can also serve as an easy to operate switch for periodic isolation of components during system maintenance, while fuses require a separate disconnect means. "Slow blow" fuses offer a time delay feature to reduce "nuisance trips" during higher initial inrush currents usually found in motor circuits. Although there are many different types and sizes of fuses, most solar and battery bank applications require a fuse that will take a momentary high current flow without failure by use of a time delay feature, but provide an instant circuit interruption for extremely high current flows.

The most popular fuses with a time delay feature and DC ratings are the Class RK5 and the Class T fuses. The Class T fuse has bolt on type connecting lugs and is usually mounted in or directly on the battery bank. This is considered a catastrophic fuse due to its very high interrupt capacity, as it is used to prevent battery system "melt down" if a major system short circuit is not stopped by circuit breakers or other fuses in the down stream circuits. The Class T fuse is available in 110, 150, 175, 200, 250, 300, and 400 amp ratings for DC circuits.

The Class RK5 fuse has blade type ends, and unlike the bolt on Class T fuse, is less expensive and can be easily removed from its holder. It should be mounted in a disconnect box having a knife switch to disconnect power for fuse replacement. It has a very good time delay feature, and will not burn out during momentary high currents typically produced when starting electric motor driven loads like pumps and fans. The Class RK5 fuse is available in 10, 20, 30, 40, 50, 60, 100, and 200 amp ratings for DC circuits.

SWITCHING ELECTRIC POWER: As stated earlier, all electric flows do not like to be interrupted once started. AC power flows are easier to turn off since by nature, the alternating voltage is switching direction 60 times each second (50 times for European standard) which means the voltage goes to zero by itself at each cycle reversal. A DC power flow does not fluctuate and does not constantly change direction like AC power. DC wall switches and circuit breakers must be designed to break the circuit very quickly. Older spring loaded snap switches are excellent choices for DC switch applications, but newer "silent" wall switches and AC rated fuses and circuit breakers may fail when used on DC circuits. It is not unusual for an AC rated switch to "weld" itself shut when it is slowly opened and the DC current arcs across the small opening trying to maintain flow. It is also possible for high current, low voltage DC power to vaporize the metal conductor inside an AC rated fuse, then continue to pass through. This is why all DC rated fuses, switches, and circuit breakers are physically larger and of heavier construction than AC hardware, and why AC rated safety devices should never be used on DC voltage wiring circuits. Using AC rated equipment with DC circuits is also against the National Electric Code and will not be acceptable to electrical inspectors.

PRELIMINARY WIRE SIZING GUIDE: When using the wire sizing TABLE 6 and 7 in the Appendix, be sure to consider all adjustment factors addressed in this chapter which may alter the wire size recommendations for your specific application.

Wire is sized by gauge number with the larger gauge number indicating a smaller wire size. After the gauge number reads ∅, still increasing wire sizes are designated by more zeros, with 2/0, 3/0, and 4/0 including larger and larger battery cables. The same wire size will have different current carrying capacities depending on insulation type, ambient temperature, and if conduit is used. TABLE #6 in the Appendix provides maximum amperage ratings for type UF (outdoor 60°C / 140°F), USE-2 (Direct Burial 90°C / 194°F), THHW (75°C / 167°F), and THWN-2 (90°C / 194°F) copper wire. Although other wire types are available, these are the most commonly used wire types found in solar and battery backup systems. TABLE #7 gives the maximum wire lengths for different wire sizes and currents.

CHAPTER XVI
CLOSING THOUGHTS

The majority of you have read this book to learn how to install a system to reduce your dependence on the utility grid, or to help you through power outages. If you are in this category, you probably have not considered the environmental impact of your decision. Just for the record, I would like to leave you with some statistics on how the environment would benefit from your energy reduction efforts. Whenever you reduce your grid electrical usage by installing more energy efficient lighting and appliances or by installing a solar or wind powered energy system, you have reduced power plant carbon dioxide (CO_2) emissions by 1 ½ pounds, sulfur dioxide (SO_2) emissions by 5.8 grams, and nitrous oxide (NOx) by 2.5 grams for every kilowatt hour you save. To put this in perspective, a single 100 watt light bulb burning 12 hours per day will consume 438 kWh per year. This produces 657 pounds of CO_2, 5.6 pounds of SO_2, and 2.5 pounds of NOx emissions at the power plant.

The Million Roof Solar Initiative recently passed by congress encourages builders to build one million homes with solar photovoltaic systems by the year 2010. Some electrical utilities facing high summer peak system demands for residential air conditioning, are already considering purchasing and installing grid tied solar arrays for residential customers that already have correctly oriented roofs. This provides the utility with the maximum solar array output, at the same time of day most electrical cooling loads peak. In many cases, these grid connected solar arrays will not include a battery bank since 100% of the power generated can be immediately used by the grid.

More and more state legislatures are passing "green power" regulations which require the local utility to purchase a small percentage of the electricity from renewable energy sources. This electricity is then sold at a higher price to those customers willing to pay a premium for this green power as a matter of conscience or stewardship. Some states do not have an adequate base of independent renewable energy suppliers established and the utility companies in these states have been paying in excess of 50¢ per kWh to solar and wind system owners in other states to help meet these local requirements. If your local utility allows transferring excess electricity back into their grid system for resale elsewhere, this can be a significant cost incentive to install a larger system. Due to the added interconnect equipment that may be required, this is usually not worth the effort for systems under 2 kW in size.

I know the concept of your local utility company getting into the act and "selling sunshine" may run contrary to your desire to see our nation reduce its dependence on a centralized power distribution system, but in some areas of the world this may be cost effective; especially for homeowners given the opportunity to lease or purchase a relatively expensive solar system on a time payment plan with the cost distributed over 10-years of monthly electric bills. It also allows small communities to be built in remote locations not served by a conventional utility line. Throughout this book we have shown only residential sized equipment. Figures 16 - 1, 16 -2, and 16 - 3 provide a glimpse of the type of commercial equipment now available to build a central community sized alternative energy system.

Emergency generators and alternative power systems in this size range usually include a duplicate of each major piece of equipment to allow shutting off and servicing any problem equipment without turning off the entire power system.

Figure 16 - 1 Large commercial DC to AC central plant battery inverters.

By setting up a private grid system and installing a centrally located large solar array, several homeowners can pool their resources and have electricity in very remote areas. Being able to lease the entire solar system and having a dependable source of electricity at a reasonable rate with others assisting with system and battery maintenance will allow small communities of like-minded families to be established where not economically possible before.

Figure 16 - 2 Central plant stacked gel cell batteries.

A dedicated power generation system for more remote parts of the country also serves to decentralize power production and reduce reliability risks associated with supplying everyone with only one interconnected grid system. Several residents planning to build in an area not served by the utility grid may want to consider forming their own electric cooperative. A centralized solar system including a large backup propane or diesel fueled industrial generator could be built and jointly owned. This may also have some tax advantages and would require much lower initial cash outlays per homeowner.

A system this size would need to be located in its own building and the commercial sized solar array could be located on a nearby south sloping hillside away from trees and other obstructions. The power would then be feed to individual homes using underground power wiring.

Figure 16 - 3 Central plant generator room and dual 500 kW generators.

With each homeowner paying for their metered usage, plant equipment costs, maintenance expenses, and operating costs could become a business expense to the owners of the decentralized grid system if separately owned and operated for profit. Who knows, our future may include thousands of privately owned solar powered utility grids dedicated to remote communities like the small self contained western towns of our early pioneer days. They say history does repeat itself, only this time these isolated pioneer towns will have electric lights, microwave ovens, satellite receivers, and whirlpools.

I have purposely kept all technical discussions to a minimum in presenting the material in this text, as I feel this introduction to off-grid and generator based power systems may be the first book the reader studies on the subject. I also believe there is a long educational process each reader will go through before actually ordering solar modules, batteries, inverters, and generators.

A thorough understanding of the basics must be absorbed first, along with an understanding of any life style changes that may be required. I sincerely hope that after reading this material, you will know what type of system will meet your needs, what loads should and should not be on the system, and that you will have a rough estimate of component sizes that will be needed. This will also allow developing a preliminary cost budget and give you a better understanding of the terms before talking with suppliers and installers.

There are many variables that must be considered to make an alternative energy system reliable, cost effective, and safe. When you are ready for actual system design and component sizing, please seek the assistance of firms trained and experienced in this specialized field of engineering, and always hire a licensed electrician for the electrical installation. What may be viewed as an added expense now, will almost always save in total system cost by avoiding poor system performance and code interpretation disputes with local building inspectors.

APPENDIX

HELPFUL CONVERSION FACTORS

The following conversion factors are related to solar heating, solar photovoltaic systems, generators, and propane and wood fuel. These will be helpful to estimate preliminary system performance.

MEASUREMENTS

1 ft.	= 0.305 meter
1 meter	= 3.28 ft.
1 watt	= 1 joule/sec
1 HP	= 745.7 watts
1 kW	= 1.34 HP
1 $ft.^3$	= 7.48 gallons (U.S.)
1 gal.	= 8.33 lbs. (water)
1 $ft.^3$	= 62.35 lbs. (water)
1 gal.	= 3.7854 liter
1 liter	= 0.264 gal.
1 $ft.^2$	= 0.0929 square meter
1 sq. meter	= 10.8 $ft.^2$
1 psi	= 2.31 ft. head (water)
1 acre	= 43,560 $ft.^2$
1 barrel	= 42 gallon (oil)
1 cord	= 4 ft. x 8 ft. x 4 ft. high

HEAT CONVERSION

1 ton	= 12,000 BTU (cooling)
1 kWh	= 3,413 BTU (heat)
1 gal.(propane)	= 92,000 BTU (heat)
1 gal.(fuel oil)	= 40,000 BTU (average)
1 cord (wood)	=30,000,000 BTU (average)

WATER USAGE

1 person	= 75gal./day(total,including washing)
1 person	= 20 gal./day (hot water only)
1 bath	= 30 gal.
1 toilet	= 3 gal./flush
1 toilet (water saver)	= 1.6 gal./flush
1 shower	= 3 to 5 gal./min.
1 shower(water saver)	= 2 gal./min.
1 dishwasher	= 7 gal./load
1 clothes washer	= 30 gal./load
1 lawn sprinkler	= 120 gal./hr.
1 horse or cow	= 12 gal./day (drinking)

PROPANE USAGE (AVERAGE)

1 gal.	= 92,000 BTU
1 gal.	= 36.39 $ft.^3$ gas
gas stove/oven	= 40 $ft.^3$/hr. gas
6.5 kW generator @ 75% load	= 47 $ft.^3$/hr. gas
500 gal. tank	=150 $ft.^3$/hr. max. delivery

TEMPERATURE CONVERSION

(° C x 9/5) + 32 = ° F

(° F - 32) x 5/9 = ° C

"RULE OF THUMB" GUIDELINES FOR SYSTEM DESIGN

- One 6 volt golf cart battery will usually store about 1 kWh of useful output energy, and one 6 volt L-16 battery will usually store 2 kWh.

- One 50 watt solar module will usually charge one 6 volt golf cart battery in one day with a fairly clear sky in the continental United States. Larger systems may require up to two batteries per module to provide longer backup time before recharging.

- Allow a minimum of four square feet of roof area for each 50 watt solar photovoltaic array capacity which equals 12.5 watts per square foot. Larger solar photovoltaic modules are which will reduce the number of individual module wiring connections and mounting hardware, but increase module weight and mounting frame load requirements.

- Although there will always be an ideal mounting position for any solar array that will produce the highest yearly performance, reasonable performance can be obtained by facing a solar array south, with a tilt angle equal or greater than the latitude of the site. A steeper tilt will increase winter energy collection, but decrease summer energy collection.

- Facing south-east improves early morning energy collection but reduces afternoon collection. The reverse is true for a south-west orientation. Vertical south facing wall mounted solar arrays are acceptable in very northern latitudes if roof mounting is not possible.

- Generators that operate at 1800 RPM instead of 3600 RPM, and operate on propane last longer and are better suited for long term backup service. Diesel generators are also OK.

- A 12 volt 2,500 watt inverter, and four to eight golf cart batteries should be adequate for a small cabin with no major appliances or well pump. A 24 volt 4,000 watt inverter and eight to sixteen L-16 batteries should be adequate for a retirement home or residence having high efficiency appliances, a well pump, but no air conditioning.

- Space heating if required, should be a combination of a wood stove and/or a propane or oil fired heater. Do not power electric heaters from a solar system. There are other more cost effective ways to heat. Quality slow speed 5 bladed ceiling fans use minimum electricity, are quiet, and should be considered as a solution to reduce air conditioning needs.

- Avoid any heating system that requires a central fan. A hydronic baseboard radiator or radiant floor heating system requires minimum electricity to operate the circulating pump and boiler. Be sure all controls are simple and do not use sensitive electronic components.

- Almost all lighting should be fluorescent, with halogen bulbs used for recessed lighting and fixtures on dimmer circuits. Avoid fixtures requiring decorative exposed bulbs.

- Use motion sensor light switches for bathroom, kitchen, corridor, and any outside lighting when possible to reduce operating time of lights in un-occupied areas.

- A packaged solar hot water heater can reduce the need to heat water with gas or electricity. Any required circulating pump and controls will need minimum electricity to operate.

- Hand tool and cellular phone battery chargers, X-10 remote control devices, and most laser printers and photocopy machines can be damaged when powered from a modified square wave inverter (low cost). Only use a pure sine wave inverter to power these loads.

- Batteries should be located in conditioned rooms or insulated battery boxes, and vented to the outdoors. The ideal battery temperature is 77° F, with 60° F to 90° F acceptable without derating charge capacity.

- Piping and ductwork should be routed to minimize elbows and changes in direction which significantly reduces pump and fan energy.

- Roof mounted solar arrays required an approved ground fault device, ground mounted arrays do not.

- Never use AC rated circuit breakers and wall switches on DC circuits. These do not meet code and these devices will quickly fail.

- When all else fails, always have several flashlights nearby!

TABLE: #1 ELECTRICAL LOADS BY TIME PERIODS

ROOM ______________________________

Quantity	Light or Appliance	Watts Each	Watts Total	5 AM - 9 AM Hours On	9 AM - 5 PM Hours On	5 PM - 10 PM Hours On	10 PM - 5 AM Hours On
	Sub-Total	-	-	-	-	-	-

TABLE: #2 WATT - HOURS BY TIME PERIOD

ROOM ______________________________

Light or Appliance	5 AM - 9 AM Watt Hours	9 AM - 5 PM Watt Hours	5 PM - 10 PM Watt Hours	10 PM - 5 AM Watt Hours
Sub-Totals				
	Period #1	Period #2	Period #3	Period #4

Select which period is highest. Add all columns together to obtain total watt-hours.

TABLE: #3 APPLIANCE ELECTRICAL LOADS

APPLIANCE **Kitchen and Miscellaneous**	**WATTS (Run)**	**WATTS (Standby)**	**kWh per YEAR**
Medium Microwave Oven w/electronic clock	1,260	3	190
Small Microwave Oven w/clock	1032	2	-
Small Microwave Oven w/Manual Timer	1030	0	-
Cappuccino Machine	758	0	-
Food Processor/Blender	151	0	-
Toaster	1,055	0	39
Dishwasher	1,250	0	700
Coffee Maker	850	0	106
Blender/Mixer	129	0	15
Hand Mixer	80	0	860
Kitchen Exhaust Hood	175	0	660
Bathroom Fan	105	0	-
Clothes Washer	500	0	103
Clothes Dryer (Propane) w/Manual Timer	277	0	-
Vacuum Cleaner	887	0	46
Curling Iron	40	0	-
Hair Dryer	745	0	14
Water Bed Heater	153*	0	-
Electric Blanket for Bed	180*	0	-
Heating Pad	65*	0	-

* Will cycle on and off to satisfy thermostat. Assume 15 minutes per hour operation. Any appliances in TABLE #3 that do not have a kWh per year value, will require estimating your own operating hours per day or week, times 365 days per year or 52 weeks per year. Any appliances given a yearly total only, must be divided into a day or week total.

TABLE: #3 continued -

APPLIANCE Heating and Refrigeration	WATTS (Run)	WATTS (Standby)	kWh per YEAR
Refrigerator/Top Freezer 14 ft. [3]	-	-	526
Refrigerator/Top Freezer 17 ft. [3]	-	-	533
Refrigerator/Top Freezer 19 ft. [3]	-	-	655
Refrigerator/Side Freezer 21 ft. [3]	-	-	760
Top Load Freezer 10 ft. [3]	-	-	349
Top Load Freezer 25 ft. [3]	-	-	570
Small Apartment Refrigerator 3.7 ft [3]	-	-	345
Bar Refrigerator 2.5 ft. [3]	-	-	303
Window AC 6200 BTU	660	0	516
Window AC 8000 BTU	890	0	667
Window AC 12000 BTU	1300	0	998
Room Dehumidifier	530	0	-
5 Blade Paddle Ceiling Fan	60	0	-
Wood Stove Blower Fan	90	0	-
Gas Furnace w/1/2 HP. Fan	909	0	-
Gas Furnace w/1/3 HP Fan	667	0	-
Gas Fired Boiler w/blower fan	155	0	-
Oil Fired Boiler w/blower fan	263	0	-
Energy saver gas DHW tank w/flue fan	144*	2	-
Doorbell transformer	2	2	18
Furnace thermostat control	2	2	18
Burglar alarm w/digital keypad	11	11	96
7 GPM ultraviolet water purifier	70	**	-

* Only when burner and fan cycles on.

** Will cycle on when water is flowing.

TABLE: #3 continued -

APPLIANCE Small Business Equipment	WATTS (Run)	WATTS (Standby)	kWh per YEAR
Halogen Desk Light w/wall cube	25	1	-
1 Tube Fluorescent Aquarium Light	20	0	-
Treadmill at medium speed w/electronic controls	280	9	-
Cordless Phone w/ wall cube	2	2	-
Credit Card Keypad Terminal	7	7	61.3
Receipt Printer for Credit Terminal	4	4	35.0
Fax Machine	20	5	-
Answering Machine	5	2	-
Ink Jet Printer (color)	17	4	-
Dot Matrix Printer	24	12	-
Laser Printer	348	8***	-
286 Computer w/ 13" Monochrome Monitor	74	0	-
166MHZ 486 Pentium w/15" Color Monitor	113	0	-
66 MHZ Computer w/15" Color Monitor	93	0-	
36" Drafting 8 Pen CAD Drum Plotter	148	0	-
Flat Bed Color Scanner	19	2	-
Photo Copier (3 copies/min.)	700	3	-
Photo Copier (10 copies/min.)	1,100	5**	-
Exit Light w/incandescent 2 @ 20 watt	40	40	350
Exit Light w/ L.E.D. Lamps	7	7	61

NOTE: *Heating element load shown is not constant due to thermostat cycling. Use 1/3 of value shown as estimate of a constant load.

** 390 watt running after warm up

*** A 348 watt heater wire will cycle on for five seconds every 20 seconds.

TABLE: #3 continued -

APPLIANCE Lighting and Audio-Video	WATTS (Run)	WATTS (Standby)	kWh per YEAR
Fluorescent Fixture 2 foot 1 @ T-12 Standard Ballast	24	0	-
Fluorescent Fixture 2 foot 1 @ T-8 Electronic Ballast	16	0	-
Fluorescent Fixture 4 foot 1 @ T-12 Standard Ballast	48	0	-
Fluorescent Fixture 4 foot 1 @ T-8 Electronic Ballast	31	0	-
Fluorescent Fixture 4 foot 2 @ T-12 Standard Ballast	95	0	-
Fluorescent Fixture 4 foot 2 @ T-8 Electronic Ballast	62	0	-
7 watt Compact Fluorescent	11	0	-
13 watt Compact Fluorescent	17	0	-
20 watt Compact Fluorescent	27	0	-
26 watt Compact Fluorescent	34	0	-
12" Color Television w/wall cube	60	2	-
19" Color Television with remote control	75	4	-
50" Projection Television	180	2	-
12" Black & White Television w/wall cube	20	2	-
Sony DSS Satellite Receiver	15	14	-
Basic Playback VCR	19	6	-
Stereo Amplifier 300 watt w/remote	75 *	3	-
CD Player w/remote	5	3	-
Cassette Tape Deck w/remote	6	1	-
Laser Disk Player w/remote	6	3	-
Stereo VCR w/remote	22	7	-

* Most stereo amplifier nameplate ratings are at maximum possible loads that can only be achieved at ear shattering volume. My tests measured 32 watts of electrical load for a 300 watt rated stereo amplifier at normal room volume settings and using quality speakers.

TABLE: #3 continued -

APPLIANCE Power Tools and Miscellaneous	WATTS (Run)	WATTS (Standby)	kWh per YEAR
Craft Glue Gun	20	0	-
Soldering Iron (100-140 watt)	125	0	-
Circular Saw (7-1/4 in.)	900	0	-
1/2 in. Drill	400	0	-
Well Pump - 1/3 HP	408	0	-
Well Pump - 1/2 HP	535	0	-
Well Pump - 3/4" HP	760	0	-
Hydronic Heating Pump - 1/20 HP	74	0	-
Hydronic Heating Pump - 1/15 HP	98	0	-
Halogen Lamps w/wall cube	Lamp Rating	1	-
Incandescent Lamps	Lamp Rating	0	-
X-10 Remote Control Modules	0.5	0.5	4.4
Shop Vac 10 Gal.	699	0	-
Sewing Machine	75	0	-
Electric Piano w/internal speaker	11	0	-
25 Gal. Aquarium Air Pump and Heater	125 *	2	-

TABLE #3 data was developed from my actual tests of appliances manufactured after 1996, using a calibrated digital kWh test meter. The metered data in these tables may differ significantly from tables in other texts which sometimes rely on printed nameplate data which usually indicates a maximum power supply capacity, not normal run wattage. Some tables found in other sources may have been based on estimates or testing less efficient appliances manufactured before 1996. I purposely tested the newer appliances to match what you will probably find today to purchase for your home. Older, less efficient appliances having an actual energy usage higher than these minimums should be avoided.

TABLE: #4 FRICTION LOSS PER 100 FEET OF PVC PLASTIC WATER PIPE

TABLE # 4 will assist you in deciding what size water line to use from your well or creek to the house. Pumping pressure is usually given in "pumping head" in units of feet, not psi pressure. For each foot of elevation, there is a pressure loss of 0.433 psi per foot rise that must be supplied by the pump in addition to the friction loss of the pipe. This means 1 psi equals 2.31 feet of elevation.

Gal./ Per Min.	Gal./ Per Hour	Head Loss Feet	Pressure Loss (PSI)	Head Loss Feet	Pressure Loss (PSI)	Head Loss Feet	Pressure Loss (PSI)	Head Loss Feet	Pressure Loss (PSI)	Head Loss Feet	Pressure Loss (PSI)
		½"		3/4"		1"		1-1/4"		1-1/2"	
1	60	1.38	0.60	.036	0.16	.011	0.05				
2	120	4.38	2.10	1.21	0.53	0.38	0.16	0.10	0.04		
3	180	9.96	4.33	2.51	1.09	0.77	0.34	0.21	0.09	0.10	0.04
4	240	17.1	7.42	4.21	1.83	1.30	0.57	0.35	0.15	0.16	0.07
5	300	25.8	11.2	6.33	2.75	1.92	0.84	0.51	0.22	0.24	0.10
6	360	36.3	15.8	8.83	3.84	2.69	1.17	0.71	0.31	0.33	0.15

Example: You have just drilled a 6" diameter well 200 feet deep. You measure down 30 feet to the water level and 150 feet down to the pump. When the pump is running the water level in the well drops another 10 feet during a pump test. Your house is on a hill 20 feet above the well location, and 400 feet from the well. You want to pump 4 gallons/minute using a 1" pipe. What are the pumping loads?

$$\text{Length of pipe} = 150 \text{ ft.} + 400 \text{ ft.} = 550 \text{ ft.}$$

Using the TABLE #4, at 4 gal./min., a 1" pipe has a 1.3 ft. loss per 100 ft.

$$\text{Pipe loss} = \frac{550}{100}(1.30) = 7.15 \text{ ft. head loss}$$

$$\text{Elevation head loss} = 30 \text{ ft.} + 10 \text{ ft.} + 20 \text{ ft.} = 60 \text{ ft.}$$

Your pump must be able to pump against these pressure losses, plus the additional load of pressuring the expansion tank which is usually 50 psi.

$$\text{System pressurization} = (50 \text{ psi})(2.31) = 115.5 \text{ ft.}$$

$$\text{Total piping head} = 7.15 + 60 + 1115.5 + 182.65 \text{ ft.}$$

For this example, select a pump that can pump 4 GPM at a 185 ft. minimum head pressure.

As stated earlier, also select a super efficient pump designed for "soft starting" and 120 volt AC operation to reduce high in-rush currents on a generator or inverter based power supply.

TABLE: #5 BATTERY STATE OF CHARGE (VOLTS)

% Full Charge	Hydrometer Reading @ 77° F	12 Volt Bank	24 Volt Bank	48 Volt Bank
100%	1.266	12.64	25.27	50.54
95%	1.259	12.59	25.18	50.36
90%	1.251	12.55	25.09	50.18
85%	1.244	12.50	25.00	50.00
80%	1.236	12.46	24.91	49.82
75%	1.229	12.41	24.82	49.64
70%	1.221	12.37	24.73	49.46
65%	1.214	12.32	24.64	49.28
60%	1.206	12.28	24.55	49.10
Maximum Daily Depth of Discharge				
55%	1.199	12.23	24.46	48.92
50%	1.191	12.19	24.37	48.74
45%	1.184	12.14	24.28	48.56
40%	1.176	12.10	24.19	48.38
Never Exceed Depth of Discharge				
35%	1.169	12.05	24.10	48.20
30%	1.161	12.01	24.01	48.02
25%	1.154	11.96	23.92	47.83

The above TABLE is for battery cells at 77° F (27° C) and measured when not under load or charging.

Repeated discharge below maximum levels indicated will significantly shorten battery life. In high ambient temperature locations, a more diluted electrolyte may be required and different hydrometer readings will result.

The following temperature corrections should be applied to Table #5 voltage readings when ambient temperature is above or below 77° F.	**Ambient Temperature**	**12 Volt**	**24 Volt**	**48 Volt**
	95° F / 32° C	- 0.3 v	- 0.6 v	- 1.2 v
	85° F / 29° C	- 0.1 v	- 0.3 v	- 0.5 v
	65° F / 18° C	+ 0.2 v	+ 0.4 v	+ 0.8v
	55° F / 10° C	+ 0.4 v	+ 0.7 v	+ 1.5 v
	45° F / 7° C	+ 0.5 v	+ 1.1 v	+ 2.1 v
	35° F / 2° C	+ 0.7 v	+ 1.4 v	+ 2.8 v

TABLE: #6 MAXIMUM WIRE AMPERAGE RECOMMENDATIONS

Wire Gauge	Battery & Inverter Wiring 30°C (86°F)Ambient Not In Conduit THHW	Battery & Inverter Wiring 30°C (86°F) Ambient In Conduit THHW	Underground Direct Burial 30°C (86°F) Ambient UF	Exterior Solar 80°C (176°F) Ambient In Conduit THWN-2, USE -2
#14 Gauge	15	15	15	6
#12 Gauge	20	20	20	8
#10 Gauge	30	30	30	12
# 8 Gauge	70	50	40	22
# 6 Gauge	95	65	55	30
# 4 Gauge	125	85	70	38
# 2 Gauge	170	115	95	53
# 1 Gauge	195	130	110	61
# 1/0	230	150	125	69
# 2/0	265	175	145	79
# 3/0	310	200	165	92
# 4/0	360	230	195	106

If physical wire size limitations or this TABLE do not provide the amperage needed for a given wire size, refer to the National Electric Code TABLE 310 - 16 through 19 for higher amperage ratings available for a given wire size when used with a higher temperature insulation.

Example: What wire size is required to connect four (4) 12 volt solar modules wired in series, to a charge controller 40 feet away? Each panel is rated at 5.6 amps at 12 volt load, and 6.3 amp short circuit.

(6.3) (1.56) = 9.83 amps at 48 volts.

From TABLE #7, a #10 wire will carry 10 amps up to 45 feet at 48 volt. Checking TABLE #6, a #10 type USE-2 wire is approved for up to 12 amps at 176°F, which is acceptable for our application.

TABLE: #7 2% WIRE LOSS TABLE

	AMPS	12 Gauge	10 Gauge	8 Gauge	6 Gauge	4 Gauge	2 Gauge	1/0	2/0	4/0
12 VOLT	5	14	22	36	57	91	144	230	290	461
	10		11	18	28	45	72	115	145	230
	15			12	19	30	48	76	96	153
	20				14	22	36	57	72	115
	40					11	18	28	36	57
	60						12	19	24	38
	100							11	14	23
	175									13
	250									
24 VOLT	5	28	45	72	114	182	289	460	580	923
	10	14	22	36	57	91	144	230	290	461
	15		15	24	38	60	96	153	193	307
	20		11	18	28	45	72	115	145	230
	40				14	22	36	57	72	115
	60					15	24	38	48	76
	100						14	23	29	46
	175							13	16	26
	250									18
48 VOLT	5	56	90	144	229	364	579	921	1161	
	10	28	45	72	114	182	289	460	580	923
	15	19	30	48	76	121	193	307	387	615
	20		22	36	57	91	144	230	290	461
	40		18	18	28	45	72	115	145	230
	60		15	12	19	30	48	76	96	153
	100					18	28	46	58	92
	175							26	33	52
	250								23	36

TABLE #7 lists the maximum recommended lengths of wire run for various combinations of wire size, current, and voltage, while maintaining a 2% voltage drop. You should only count the actual one way distance between wiring components, since the resistance for the total length of circuit wire out and back the electricity must travel to complete the circuit is accounted for in these table values.

TABLE: #8 BATTERY BANK SIZES AND WIRING

NOTE: Overall dimensions shown include 1/2" space between batteries. Battery dimensions may vary slightly between manufacturers.

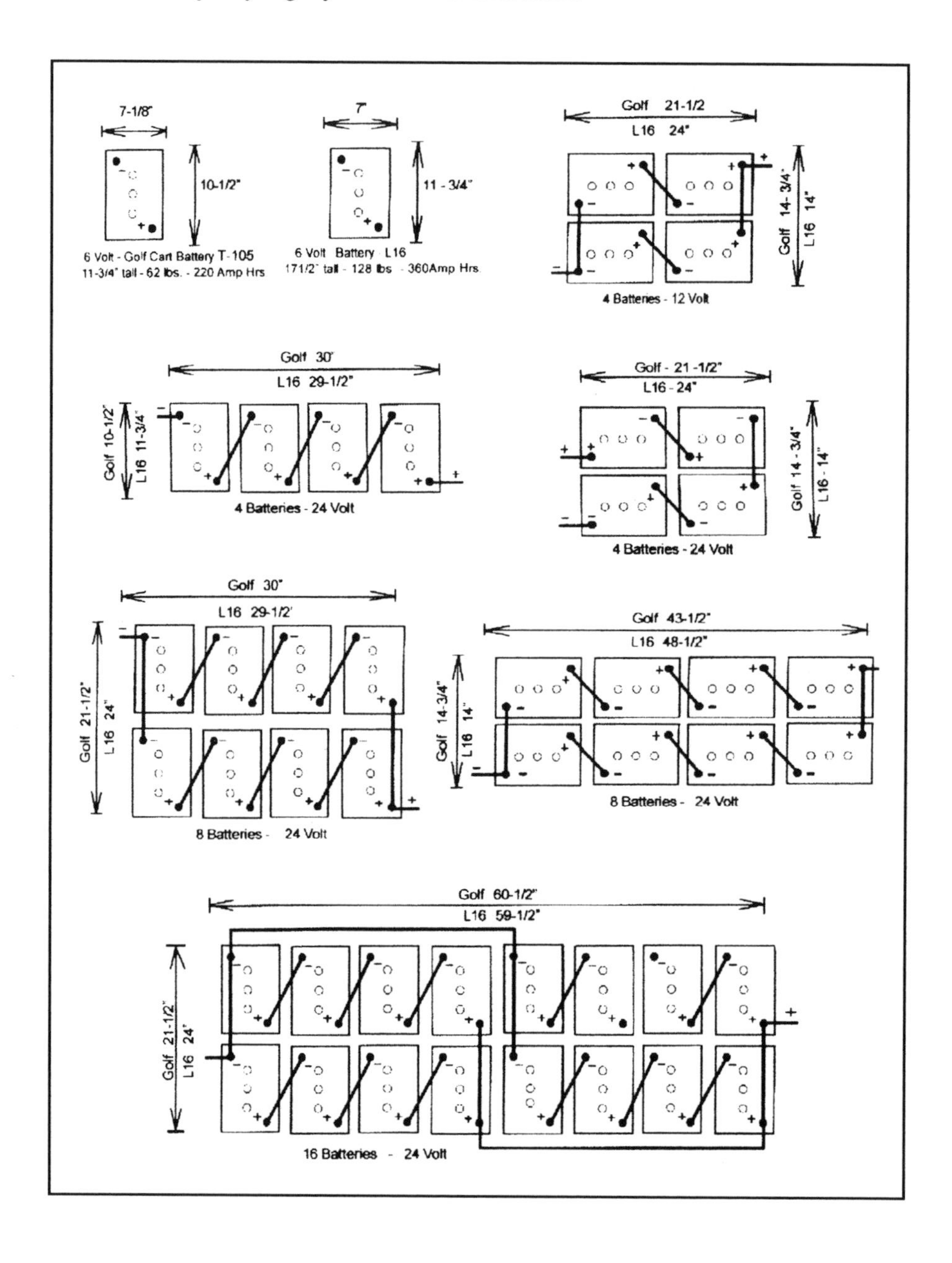

TABLE: #9 DIMENSIONS of SOLAR PV PANELS (Inches)

Manufacturer	MODEL #	LENGTH	WIDTH	WEIGHT	AMPS	VOLTS	WATTS
Astro Power	AP1206	58.1	24.3	26.1	7.1	16.9	120
British Petroleum	BP275	46.8	20.9	20	4.45	17.0	75
British Petroleum	BP590	46.8	20.9	20	4.86	18.5	90
Kyocera	KC40	25.7	20.7	16	2.34	16.9	40
Kyocera	KC60	29.6	25.7	20	3.55	16.9	60
Kyocera	KC80	38.4	25.7	25	4.73	16.9	80
Kyocera	KC120	56.1	25.7	30	7.10	16.9	120
Solavolt	SV7500	47.2	18.9	16.5	4.17	17.3	73
Solavolt	SV8500	47.2	20.9	18.7	4.91	17.4	85
Solarex	MSX-60	44	20	16	3.50	17.1	60
Solarex	VLX-80	43.5	26	21	4.71	17.0	80
Solarex	MSX-120	43.8	39	30.8	7.00	17.1	120
Siemens	SM46	42.7	13	10.2	3.15	14.6	46
Siemens	SM55	50.9	13	12	3.15	17.4	55
Siemens	SP75	47.3	20.8	16.7	4.40	17.0	75
Siemens	SM100	51.8	25.9	25.1	5.90	17.0	100
Siemens	SR100	59.0	23.4	24.0	5.90	17.0	100
Siemens	SM110	51.8	25.9	25.1	6.3	17.5	110
Unisolar *	US42	53.8	20	21	2.70	15.6	42
Unisolar *	US64	53.8	29	30	4.10	15.6	64
Solec/Sanyo	S55	50.8	13	13.2	3.22	17.1	55
Solec/Sanyo	S100	51	25	23.4	5.88	17.1	100

TABLE NOTES: "Volts" and "Amps" columns are under load at rated maximum power.
"Watts" column is wattage at rated maximum power.
Weight column is given in pounds.
* Unisolar panels have non-glass construction

TABLE: #10 SIMPLIFIED METHOD TO PREDICT SOLAR ARRAY PERFORMANCE

This methodology assumes you have already selected the number of photovoltaic modules your budget will allow. It does not calculate the required array size. You will need to know the wattage rating for the PV module you are considering. Keep in mind that these manufacture ratings are based on electrical output under very ideal conditions with the module oriented directly towards the sun with no clouds. These maximum solar module ratings can be adjusted for your location and mounting as follows:

A.	Enter nameplate rating of panel (watts)	A	
B.	Enter Quantity of panels in array	B	
C.	Maximum array output (A x B)	C	
D.	Adjust for location ($Factor_1$ x C)	D	

$Factor_1$

1.00	Southern states and warm climates
0.95	Mid-latitude states with sunny winters
0.90	Mid-latitude states with partly cloudy winters
0.85	Northern or mountain states with partly cloudy winters
0.80	Northern or mountain states with cloudy winters
0.75	Any location with cloudy winters and summers

E.	Adjust for tilt and orientation ($Factor_2$ x D)	E	

$Factor_2$

1.00	Sun tracking mount
0.90	South facing with seasonal tilt change
0.80	South facing with tilt fixed at latitude
0.78	South facing with tilt fixed at latitude ± 15 degrees
0.70	Off south and fixed tilt at latitude
0.60	Vertical south wall mounting

F. Adjust for type of panel F [ ]

Factor $_3$

1.00 Single crystal cells
0.98 Polycrystalline cells with seasonal tilt
0.96 Polycrystalline cells with fixed tilt
0.60 Amorphous cells with fixed tilt

G. Adjust for battery and inverter G [ ]
(Factor $_4$ x F)

Factor $_4$

1.00 Array output directly used by DC loads without inverter
0.95 Array output directly used by AC loads using inverter
0.85 Array output charges battery and battery supplies DC loads
0.80 Array output charges battery and inverter supplies AC loads

H. Calculate expected daily watt hours harvested H [ ]
(Factor $_5$ x H) Watt-Hours

Factor $_5$

6.0 Locations with very sunny days and full sun from 9:00 AM until 3:00 PM
5.5 Locations with mostly sunny days and full sun from 9:00 AM until 3:00 PM
5.0 Locations with partly sunny days and full sun from 9:30 AM until 2:30 PM
4.0 Locations with cloudy days and some sun from 10:00 AM until 2:00 PM

Compare the estimated daily watt hours harvested (H) with the daily watt hour loads that you calculated using TABLE #2 in the Appendix. This method will give you an idea of the yearly average your solar system will provide for your electrical load.

$$\frac{\text{Daily Energy Harvested}}{\text{Daily Energy Consumed}} = \square$$

Although this may appear to be a very simplified analysis, even the most sophisticated computer solar analysis methods still rely on regional weather data files which rarely match the weather and solar insolation for your location. In addition, all weather files are an average of up to 20 years of data and a specific year will never exactly match this data.

Example: You are buying an 8 panel photovoltaic kit consisting of 60 watt polycrystalline panels mounted on a fixed rack which is tilted at latitude and south facing. This system will include 8 golf cart batteries and a 1,500 watt inverter. All loads are 120 volt AC. The site is an off-grid cabin near Atlanta, Georgia. Your appliances and lights add up to 3,800 watt-hours per day. What is the expected average watt hours harvested per day and the yearly system performance?

TABLE: #10 EXAMPLE

SIMPLIFIED METHOD TO PREDICT SOLAR ARRAY PERFORMANCE

This methodology assumes you have already selected the number of photovoltaic modules your budget will allow. It does not calculate the required array size. You will need to know the wattage rating for the PV module you are considering. Keep in mind that these manufacture ratings are based on electrical output under very ideal conditions with the module oriented directly towards the sun with no clouds. These maximum solar module ratings can be adjusted for your location and mounting as follows:

A. Enter nameplate rating of panel (watts) — A: **60**

B. Enter Quantity of panels in array — B: **8**

C. Maximum array output (A x B) — C: **480**

D. Adjust for location ($Factor_1$ x C) — D: **432** (0.9)(480)

$Factor_1$

1.00	Southern states and warm climates
0.95	Mid-latitude states with sunny winters
0.90	Mid-latitude states with partly cloudy winters
0.85	Northern or mountain states with partly cloudy winters
0.80	Northern or mountain states with cloudy winters
0.75	Any location with cloudy winters and summers

E. Adjust for tilt and orientation ($Factor_2$ x D) — E: **345.6** (0.8)(432)

$Factor_2$

1.00	Sun tracking mount
0.90	South facing with seasonal tilt change
0.80	South facing with tilt fixed at latitude
0.78	South facing with tilt fixed at latitude ± 15 degrees
0.70	Off south and fixed tilt at latitude
0.60	Vertical south wall mounting

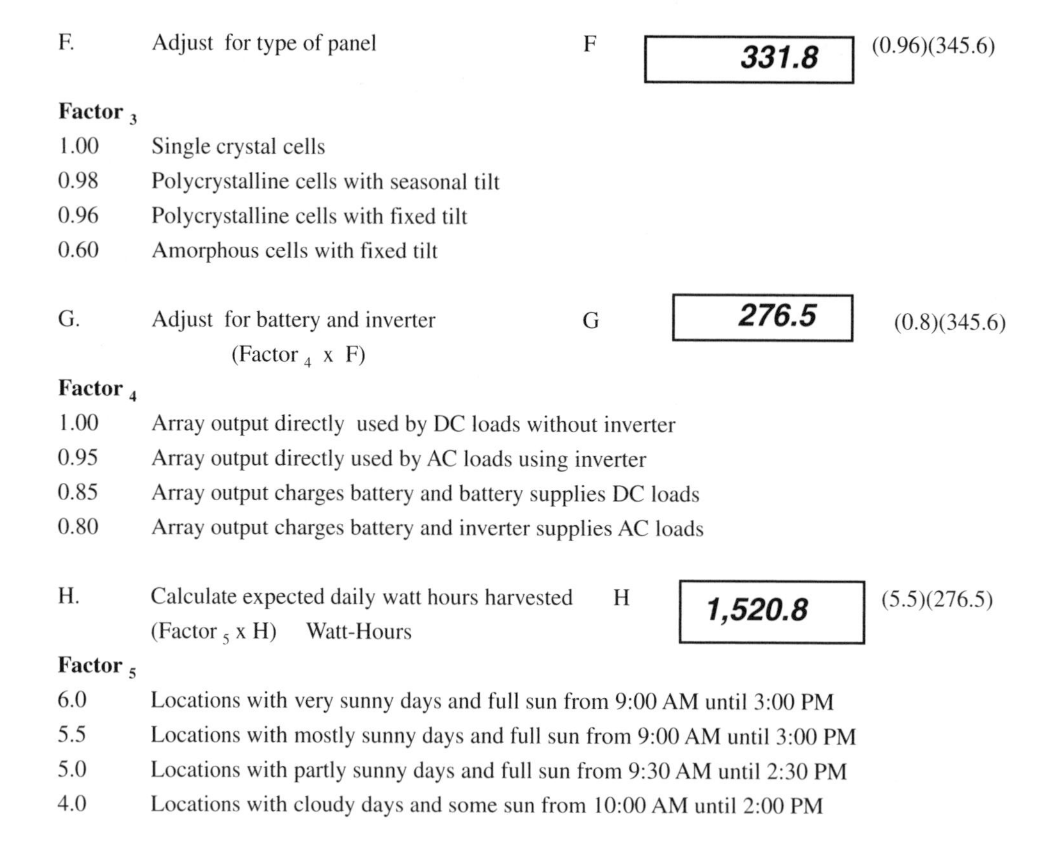

F. Adjust for type of panel — F: **331.8** (0.96)(345.6)

Factor $_3$

1.00	Single crystal cells
0.98	Polycrystalline cells with seasonal tilt
0.96	Polycrystalline cells with fixed tilt
0.60	Amorphous cells with fixed tilt

G. Adjust for battery and inverter (Factor $_4$ x F) — G: **276.5** (0.8)(345.6)

Factor $_4$

1.00	Array output directly used by DC loads without inverter
0.95	Array output directly used by AC loads using inverter
0.85	Array output charges battery and battery supplies DC loads
0.80	Array output charges battery and inverter supplies AC loads

H. Calculate expected daily watt hours harvested (Factor $_5$ x H) Watt-Hours — H: **1,520.8** (5.5)(276.5)

Factor $_5$

6.0	Locations with very sunny days and full sun from 9:00 AM until 3:00 PM
5.5	Locations with mostly sunny days and full sun from 9:00 AM until 3:00 PM
5.0	Locations with partly sunny days and full sun from 9:30 AM until 2:30 PM
4.0	Locations with cloudy days and some sun from 10:00 AM until 2:00 PM

Compare the estimated daily watt hours harvested (H) with the daily watt hour loads that you calculated using TABLE #2 in the Appendix. This method will give you an idea of the yearly average your solar system will provide for your electrical load.

$$\frac{\text{Daily Energy Harvested}}{\text{Daily Energy Consumed}} = \frac{1{,}520.8}{3{,}800.0} = 40\%$$

Although this may appear to be a very simplified analysis, even the most sophisticated computer solar analysis methods still rely on regional weather data files which rarely match the weather and solar insolation for your location. In addition, all weather files are an average of up to 20 years of data and a specific year will never exactly match this data.

Example: You are buying an 8 panel photovoltaic kit consisting of 60 watt polycrystalline panels mounted on a fixed rack which is tilted at latitude and south facing. This system will include 8 golf cart batteries and a 1,500 watt inverter. All loads are 120 volt AC. The site is an off-grid cabin near Atlanta, Georgia. Your appliances and lights add up to 3,800 watt-hours per day. What is the expected average watt hours harvested per day and the yearly system performance?

Answer: 1.5 kWh per day harvested with a 40% yearly performance.

Quick Battery Check:

Although the size of the battery bank was not considered in the above quick solar array sizing calculation, battery capacity can affect system performance if undersized or oversized. As a starting point, assume each golf cart battery can store 1 kWh of potential energy and each L-16 battery can store 2 kWh of energy when powering an AC inverter.

This means the eight golf cart batteries in our example will store 8 kWh of energy if taken to a fully discharged state. (1 kWh x 8)

If our example solar array generates 1.5 kWh per day, our battery bank will require over five days to fully recharge (1.5 x 5 days) = 7.5 kWh.. If we assume a maximum depth of discharge of 50% it will take over two days.

$$\frac{(8)(1\ kWh)(50\%)}{1.5\ kW} = 2.7\ days$$

Ideally, we would want the daily energy being taken out of a solar charged battery to be fully recharged the next day. Therefore, unless our example is an undersized weekend cabin design with the entire week to recharge, this system would probably provide better performance using four batteries instead of eight.

American Wind Energy Association
122 C Street, NW - 4th Floor
Washington, DC 20001

Phone: 202-383-2500
Web Site: www.awea.com

Appliance Efficiency Database
California Energy Commission
1516 Ninth Street
Sacramento, California 95814-2950

Phone: 916-654-4058
Web Site: www.energy.ca.gov

Backwoods Home Magazine
Post Office Box 712
Gold Beach, Oregon

Phone: 800-835-2418
Fax: 541-247-8600
Web Site: www.backwoodshome.com

Energy Efficiency Network
U. S. Department of Energy
Forrestal Building
1000 Independence Avenue, SW
Washington, DC 20585

Phone: n/a
Web Site: www.doe.gov

Home Power Magazine
Post Office Box 520
Ashland, Oregon 97520

Phone: 800-707-6585
Web Site: www.homepower.com

Solar Energy International
Post Office Box 715
Carbondale, Colorado 81623-0715

Phone: 970-963-8855
Web Site: www.sei@solarenergy.org

Solar Design and Packaged Systems

Dunimis Technology Inc.
Post Office Box 10
Gum Spring, Virginia 23065

Phone: 804-784-0063
Web Site: www.dunimis.com

INDEX

A

B

C

D

E

F

G

H

I

K

L

M

N

O

P

R

S

T

U

V

W

Y